Mountain Flying

TAB
PRACTICAL
FLYING SERIES

Other Books in the TAB PRACTICAL FLYING SERIES

Mountain Flying

*Doug Geeting and
Steve Woerner*

TAB Books
Division of McGraw-Hill, Inc.
Blue Ridge Summit, PA 17294-0850

Portions of Chapter 3 have been adapted from *Aviation Weather* by FAA/NWS.

Appendix A previous appeared, in somewhat different form, in *Alaska Flying*.

Front Cover: *Dawn light on Denali National Park, Alaska.* (Galen Rowell / Mountain Light)
Back Cover: *Co-author Doug Geeting.* (Galen Rowell / Mountain Light)
Title Page: *Skiplane, flown by Talkeetna (Alaska) pilot Cliff Hudson, about to land on
        Kahiltna Glacier.* (Galen Rowell / Mountain Light)

FIRST EDITION
FIFTH PRINTING

© 1991 by **Doug Geeting and Steve Woerner**.
Published by TAB Books.
TAB Books is a division of McGraw-Hill, Inc.

Printed in the United States of America. All rights reserved. The publisher takes no
responsibility for the use of any of the materials or methods described in this book,
nor for the products thereof.

Library of Congress Cataloging-in-Publication Data

Geeting, Doug.
    Mountain Flying / by Doug Geeting and Steve Woerner.
        p.   cm.
    Includes index.
    ISBN 0-8306-9426-9   ISBN 0-8306-2426-0 (pbk.)
    1. Mountain Flying.   I. Woerner, Steve.   II. Title.
TL711.M68G44   1988
629.132′5214—dc19   88-20142
                            CIP

To my parents, Warren and Shirley, and my brother Jim, for their encouragement through the years and believing in my dreams . . .

To Sandy and all my friends in Talkeetna, Alaska, who endured my frustrations and added moral support for the writing of this book . . .

I thank you.

D.G.

To my wife, Rosemary. Thanks for all the support.

S.W.

Keep your back door open and your stairway down and clear.

—Doug Geeting

Contents

Acknowledgments

We wish to express our sincere thanks and appreciation for special guidance and contributions of the following:

Galen and Barbara Rowell and Mountain Light Photography Studio in Albany, California, for the exceptional and unique photography they contributed.

David Whitelaw, Whitelaw Illustration and Design, Anchorage, Alaska, for his artistic talents and contributions.

And others, including Clay Lacey, Alvin S. White, Ray Arnold, Bud Wallan, Jim Bagian, M.D., Peter Hacket, M.D., Rob Roach, M.D., Bert Puskas, M.D., Larry Casarow, Al Turner, Kay Cashman, Bob Stine, Kitty Banner Seemann, Northern Lights Avionics, and Cliff Hudson for their counsel and assistance in completing this project.

Introduction

Mountain flying is the ultimate form of VFR flight. While challenging your skill and ability to perform precisely, mountain flying provides an opportunity to traverse spectacular and breathtaking terrain. There is an old saying, however, that necessarily applies here: "Increased reward is often accompanied by increased risk."

The risk of mountain flying can be minimized. Good judgment is a natural outcrop of being well prepared. Just as athletes must prepare for an event through training, so must pilots. Practicing maneuvers until they become second nature to you, thorough preflight planning, and maintaining an awareness of personal and mechanical limitations will give you the tools necessary to exercise good sound judgment in crucial situations.

Safety must be first, second, and last. Don't be the weak link in the safety chain. As with flying, but most especially with mountain flying, you must fly by the rules—and not just the FAA rules. You must be cognizant of your responsibility towards yourself and your passengers. You must employ good old-fashioned common sense.

In this book, we will examine mountain flying with respect to the pilot, the airplane, weather and topography, and mechanical performance capabilities, and will offer some personal views on certain techniques unique to mountain flying.

The information contained here comes from years of personal flying experience and a good deal of training and research. The procedures and maneuvers we will detail represent an integration of not only this research and training, but personal preferences as well. If you are an experienced mountain flier, you may indeed have different methods that serve equally well or even better. If you are less experienced, we encourage you to seek additional professional training if you plan to pursue mountain flying aggressively.

We do not profess to be the only word or the last word on this subject but rather offer these points for your consideration. We hope the ideas we present will stimulate you to develop the aggregate of knowledge and skill necessary to fly safely in the mountains.

1

The Pilot

FLYING HAS BEEN ONE OF MAN'S PERSONAL QUESTS FOR THOUSANDS OF YEARS. Here, in the twentieth century, man's dream has been realized. With flight finally possible, it seems as though all the world is within easy grasp. For those who are ready, willing, and able to accept the challenge of three-dimensional travel, previously hard-to-reach and inaccessible places are now opened up. In the mountains, especially, people can fully appreciate the expediency of this form of travel. Deep valleys and towering ranges can be crossed in a matter of minutes, which in an earlier age might have taken days or even weeks.

The accessibility of remote locations by aircraft has opened up a whole new realm of recreation. Off-pavement air camping, for example, has become very popular, especially in the mountain states and Canada. Some of the best fishing holes in the world are located high in the mountains and, more often than not, can be reached only by aircraft. Today, one of the best ways to get the upper ground on the competition is to fly into a hunting camp or special wilderness spot. Even photographers, hikers, mountain climbers, and nature lovers are discovering that flying can put them in places where few others have been.

For the pilot, mountain flying is flying at its very best. It is precision VFR flight. As such, it leaves no room for complacency or apathy. It can be terribly unforgiving for those who ignore the potential dangers. Over flatland, pilots enjoy a certain margin of safety by simply clearing all obstacles. Flying through the mountains, however, is often an exact science. Skills in pilotage and dead reckoning must be honed to a razor's edge. A pilot must possess an alert attitude and be in near-perfect health to negotiate obstacles without incident.

PILOT HEALTH AND PHYSICAL ANOMALIES

A healthy body, an alert mind, and flight proficiency are important to any flight. These same attributes become essential to flights conducted in and around large immovable obstacles, like mountains.

Except for those who fly gliders or balloons, every pilot must hold a valid medical certificate. Standards for medical certification are contained in FAR Part 67. Included are basic minimums for health and physical fitness. Examinations are required every 6, 12, or 24 months, depending on the level of responsibility you exercise as an aviator. During these intervals it is your obligation to monitor and maintain your health.

Common everyday discomforts, like headaches and colds, restrict pilot performance. The distraction of a minor illness or discomfort may impair your judgment, your ability to stay alert, and even your memory. And the medication you take to remedy an ailment can be just as harmful to performance as the infirmity itself. Tranquilizers, sedatives, pain relievers, cough-suppressants, antihistamines, blood pressure medicine, muscle relaxants—even medication to control motion sickness—all may depress the nervous system and affect the same crucial functions that the illness effects. Some drugs (the type that suppress breathing) can make pilots more susceptible to hypoxia.

OVER-THE-COUNTER DRUGS

Most pilots believe over-the-counter drugs are safe to use while flying. Nothing could be further from the truth. Over-the-counter drugs can be just as hazardous as prescription drugs. Different people react to the same drug in different ways. Mixing drugs can cause a synergistic reaction; that is, the reaction in an individual may be two, three, even ten times as great as when the same drugs are taken separately.

The Effects of Some Common Drugs

Aspirin affects regulation of body temperature by acting on the hypothalamus. It affects the acid-base balance of the body, causing a variation of the rate and depth of respiration. Two aspirin have the potential of increasing oxygen consumption and carbon dioxide production. Aspirin has a corrosive action on the stomach. For some people, two aspirin on an empty stomach can irritate the

Alaskan bush pilot Kitty Banner Seemann unties one of Doug Geeting's Cessna 185s. (Galen Rowell / Mountain Light)

lining enough to draw a teaspoon of blood. Corrosive action can also promote gastrointestinal problems.

Aspirin also reduces the clotting ability of the blood. If you are injured and have been taking aspirin, you may have higher-than-normal blood loss.

Nasal decongestants may cause prolonged increases in blood pressure, dry mouth, insomnia, fatigue, headache, dizziness, incoordination, confusion, drowsiness, amnesia, possible depression, rapid heart rate, blurry or double vision, euphoria, lassitude, tremors, nausea and vomiting, and diarrhea.

Sleeping pills depress the central nervous system and autonomic nervous system. They reduce alertness and may cause decreased coordination, drowsiness, rebound anxiety, and restlessness. The use of sleeping pills may reduce respiratory drive, thereby contributing to hypoxia.

Throat lozenges damage the blood cells. They can cause damage to the kidneys and liver and to mucous membranes. Allergic reactions, like swelling of the mouth, throat, and respiratory tract, can also occur. When taken in excess, lozenges can actually cause nerve cell damage.

Cough drops, when taken in high dosages, depress the central nervous system and reduce reaction time.

Caffeine can cause wakefulness, tremors, gastric hyperacidity and indigestion, cardiac arrhythmias, increased heart rate, headache, dizziness, and nausea. It also promotes body dehydration through increased urine output. Caffeine withdrawal may cause headache, instability, drowsiness, and lassitude.

Nicotine increases blood pressure, constricts small blood vessels, increases need for oxygen by 10–15%, and can quite possibly double your normal reaction time by paralyzing nerve cells.

ALCOHOL

Fortunately, aviation accidents involving alcohol impairment are on the decline, but they have not disappeared. According to the National Transportation Safety Board (NTSB), nearly 7% of all fatal accidents in the aviation industry are still alcohol-related. Alcohol dulls a pilot's judgment and hampers his or her ability to perform necessary functions quickly. Flying in the mountains, a pilot must stay alert and be able to react to changing situations deliberately, quickly, and accurately.

FAR 91.11(a) prohibits a pilot from taking the controls within eight hours after the consumption of alcohol. A much better rule of thumb is to wait 24 hours "from bottle to throttle."

FATIGUE

Fatigue is "gremlin maximus." A pilot must perform at his very best from the moment a flight begins, all the way through to the end when he shuts the engine off. But a pilot who feels rested and alert when he takes off from his home base is often exhausted after enduring hours of noise, vibration, a tight cockpit, congested airspace, and bad weather. When he approaches his destination airport he is worn out—at a time when he must be sharp and attentive to details.

Fatigue is a thief. It steals a pilot's ability to concentrate. The most common problem associated with fatigue is an inability to properly judge speed, altitude, or distance. Slow reaction time and inattentiveness are hallmarks of fatigue. The boredom of flying on autopilot for hours can also be fatiguing. Combined with the annoyances mentioned above, the net result can be fatal.

You can help prevent fatigue by checking your ground reference points frequently. On long flights take a break; plan for rest stops at least every three to four hours. Most conventional general aviation aircraft need refueling at these intervals anyway. Take advantage of refueling stops by stretching your legs. In flight you can help control fatigue by sitting up straight.

EFFECTS OF ALTITUDE

Pilots and mountain climbers are quick to recognize the association between hypoxia and altitude; the connection is hammered into every training program. Hypoxia can be a serious problem if not checked. In addition to your brain, altitude can affect your ears, sinuses, teeth, and gastrointestinal tract. Scuba divers who fly home after a day of deep sea exploration are much more susceptible to decompression sickness ("the bends") due to the increase in altitude.

Hypoxia

Hypoxia is the lack of sufficient oxygen in the body tissues to keep the central nervous system and other organs functioning properly. The condition is brought on by reduced atmospheric pressure; that is, the pressure that forces oxygen from the lungs into the blood supply, and then into the body tissues, is too weak. Smoking, alcohol, antihistamines, tranquilizers, sedatives, and analgesics can lower a person's tolerance to hypoxia, as can anemia, carbon monoxide, and fatigue. Most lung disease and some types of heart disease can also bring on the condition.

The body does not have an "early warning system" to let you know you're becoming hypoxic. Rather, the symptoms are insidious—they develop so gradually that you may never become aware of them. This problem is further complicated by the fact that during the first stages of hypoxia you may feel a false sense of well-being or euphoria, making it difficult to believe that there is actually anything wrong. The following symptoms are indicative of hypoxia:

⚠ a significant loss of night vision

⚠ blurred vision

5

⚠ loss of peripheral vision

⚠ anxiety

⚠ increased heart rate

⚠ headache

⚠ nausea

⚠ dizziness

⚠ slow thinking

⚠ fatigue

⚠ mental confusion

⚠ sense of well-being, euphoria, giddiness

⚠ impaired judgment

⚠ poor reaction time

⚠ poor coordination

Hypoxia can cause a pilot to lose consciousness. If left unchecked it can even cause death. The symptoms manifest themselves differently from one day to the next in the same individual and typically will be different in each person. Depending on the metabolic makeup of the individual, during daylight hours the symptoms may occur as low as 5,000 feet or not until reaching 15,000 feet. In an average healthy person, night vision may begin to be affected at 5,000 feet with as much as a 25% reduction in visual acuity at 8,000 feet. The same person may not notice any effects at these altitudes during daylight.

At 10,000 feet most people will experience fatigue and lack of precise concentration. After only a few hours pilots may be unable to perform simple calculations. More pronounced symptoms begin to occur between 12,500 and 14,000 feet. To preserve peak performance, pilots should don an oxygen mask at 5,000 feet on night flights and at 10,000 feet during the day. All pilots should know their own limitations. They should be able to recognize their own symptoms and know when they require the use of oxygen.

The average person can maintain a level of useful consciousness without supplemental oxygen as indicated in TABLE 1-1.

Pilots who fly at high altitudes every day tend to become acclimated to those altitudes. Similarly, pilots who live at high elevations are generally able to fly at higher altitudes than those who live at sea level. For example, a Denver pilot will tend to be less affected by hypoxia at 14,000 feet than a pilot who lives in Los Angeles.

Blood Donation. Because it takes at least several weeks for a person's blood supply to return to normal after donating blood, flying shortly after donation can pose a high risk of hypoxia.

Altitude (FT MSL)	Time
40,000	15 sec
35,000	20 sec
30,000	30 sec
28,000	1 min
26,000	2 min
24,000	3 min
22,000	6 min
20,000	10 min
15,000	Indefinitely

Table 1-1. *Average duration of useful consciousness without supplemental oxygen. These figures do not reflect individual tolerances for oxygen deprivation.*

(Source: FAA Office of Aviation Medicine Report AM 70-12)

Special Hypoxia Training. The U.S. Air Force offers physiological training classes to general aviation pilots at selected Air Force bases throughout the country. These classes are designed to help pilots recognize their own supplemental oxygen requirements at specific altitudes and to introduce them to the various oxygen systems. Classes are conducted in a safe, controlled environment under the supervision of professionals who are experienced in the effects of hypoxia. If you are interested in attending these classes, contact your local FAA Office and ask for the location of the nearest training facility, or write to:

Airman Education Section
AAM-142
FAA Civil Aeromedical Institute
P.O. Box 25082
Oklahoma City, OK 73125

A brief overview of the different types of oxygen systems available to the general aviation pilot is included in the next chapter.

Ears and Sinus Cavities

Every pilot has experienced the sensation of ears "popping" when changing altitude. We yawn, swallow, or hold our noses and blow to equalize the pressure. Failure to equalize the pressure can have serious consequences. More importantly, failure to *symmetrically* equalize the pressure in the middle ear on both sides may induce severe vertigo, which can make control of the aircraft virtually impossible.

Blockage in the eustachian tube can produce severe ear pain and loss of hearing that can last anywhere from several hours to several days. The eardrum may rupture in flight or even after landing. Fluid may accumulate in the middle ear and subsequently become infected. Pilots with head colds or allergic respiratory conditions are more likely than others to experience severe middle-ear problems in flight. And remember, the same applies to passengers.

Sinus cavities are likewise affected by air pressure differences while climbing or descending and can cause a fierce headache. Blocked maxillary (upper jaw) sinuses can cause the upper teeth to ache. Colds and allergies make people more susceptible to sinus problems in flight.

The Valsalva Maneuver. Pain in the ears, sinuses, or teeth detracts from a pilot's ability to perform simple tasks like making routine calculations. To unblock the ear and sinus cavities, pilots and passengers can swallow or yawn. The most popular method is called the *Valsalva maneuver*, in which the person pinches their nose and closes their mouth while blowing hard. A less proven method involves rubbing the sides of the neck where the eustachian tubes are located to help relieve the pain.

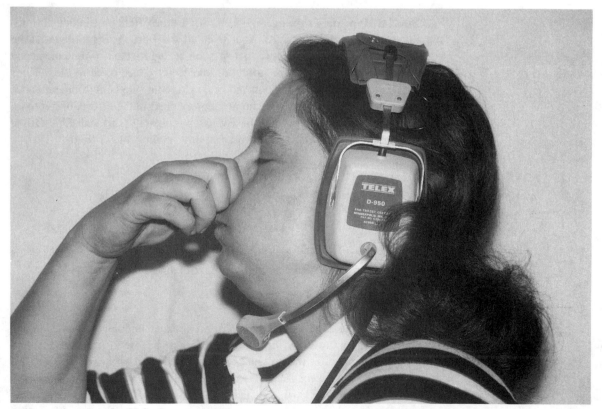

Pilot performing the Valsalva maneuver. (Steve Woerner)

Decompression Sickness

The body absorbs excess nitrogen during scuba dives. In body tissues, this gas forms bubbles that expand with altitude. These bubbles lodge in the joints of the body, and cause a painful condition known as *the bends*. The bends can occur even at low altitudes if sufficient time has not been allowed for the body to rid itself of evolved gas. The recommended waiting time before flights to cabin pressure altitudes of 8,000 feet or less is at least four hours following shallow non-controlled dives and at least 24 hours following deeper controlled dives (decompression diving). The waiting time before flight to cabin pressure altitudes above 8,000 feet should be at least 24 hours after any scuba diving.

The *aviation bends* are a variation of this decompression sickness. Unrelated to diving, the symptoms are the same and usually occur when a person ascends too quickly from a low to a high altitude. Mountain pilots often "ride" the mountain waves to climb up and over ranges. This ride is sometimes rapid and, in nonpressurized aircraft, can cause the bends in both pilot and passengers.

HYPERVENTILATION

Hyperventilation is best described as the loss of excessive amounts of carbon dioxide from the blood, resulting in a chemical unbalance in the body. Hyperventilation produces many of the same symptoms as hypoxia. The condition is brought on by overventilating the lungs by breathing too rapidly and too deeply.

In stressful situations, pilots may find that they hyperventilate rather easily. The most common example in mountain flying happens when a pilot gets excited about a stressful situation. Flying up a narrow box canyon or not having enough altitude to cross the ridge in front of you may bring on symptoms of rapid breathing, rapid pulse rate, a tingling sensation in the fingers and mouth, dizziness, disorientation, and an inability to concentrate on simple tasks.

You can bring these symptoms under control quickly by deliberately controlling your breathing. A good rule of thumb is to take a deep breath, hold, and release it at five-second intervals until the symptoms disappear. If the symptoms do not disappear in a few minutes, breathe slowly into a paper bag or an oxygen mask (without using oxygen) until symptoms go away.

Hyperventilation syndrome may occur in passengers, especially those who are not used to wearing oxygen masks. They might feel as though they are not getting enough oxygen and, therefore, breathe more deeply and rapidly than they would normally. Pilots should take care to explain the use of oxygen to their passengers before the flight so that this anxiety will not develop.

CARBON MONOXIDE POISONING

Most single-engine aircraft are heated by air that passes over the exhaust manifolds. A defective heat exchanger—one with leaks or cracks or broken seals—

can be dangerous. Exhaust fumes may enter the cabin area through the firewall or even through the landing-gear box. Cessnas have a rubber seal where the spring-gear leg is attached to the fuselage. When damaged or worn out, a gap will form in this area.

Carbon monoxide is colorless, odorless, and tasteless. It sneaks up on you slowly, and it sticks to the hemoglobin in your blood with a bond more than two hundred times stronger than oxygen. High concentrations of carbon monoxide in the aircraft cabin can make adequate oxygenation of the blood difficult or impossible. By "taking up space" in the blood, it prevents oxygen from attaching to the hemoglobin.

A deteriorated rubber seal at the top of Cessna gear legs can allow deadly carbon monoxide to seep into the cabin. (Steve Woerner)

Carbon monoxide detector. (Oleve Woerner)

Beware of Heaters

Even in the summer months, mountain pilots often find it necessary to close off the fresh air vents and turn on the heater. If you experience a headache, drowsiness, or dizziness while the heater is on, turn it off, descend to a lower altitude, and open the fresh air vents. If necessary, open the windows. If you have an oxygen system on board, put it on and set the regulator to deliver 100% oxygen. Because carbon monoxide may take anywhere from 24 to 48 hours to leave the blood, it is advisable to land immediately. Carbon monoxide is a chemical asphyxiant. It is a poisonous gas and can kill you.

Carbon monoxide detectors are available at most aviation repair stations. They cost about $2.00 and, needless to say, are worth the investment. Attach them to the side panel (near the heater vent) and the instrument panel (in front of the pilot). You need to replace the detectors every six months.

Feeling good—being healthy and alert—is an important aspect of flight. In the mountains, your senses must remain keen, and your body must be able to react quickly and decisively to potentially dangerous situations. Don't dismiss our emphasis on physical well-being. It is a real part of flying safely in the mountains. Do it in good health or don't do it at all.

ILLUSIONS AND DISORIENTATION

The eyes, ears, and skeletal muscles are the human body's sensory receptors, helping us distinguish between reality and illusion within our environment. *Illusion* occurs when the senses receive conflicting messages. Pilots become disoriented when they perceive the information seen on cockpit instruments as being different from those messages received by other senses.

The mountain environment can place a pilot in situations where illusions are likely to occur. There may be no horizon to use as a reference point. Haze and reflected light may decrease depth perception. Flights conducted above the treeline without familiar reference images (trees, cars, people, and the like) may also decrease depth perception.

Visual sensations are received via the eye. Impulses in the rods and cones travel through the optic nerve to the brain where they are processed and interpreted. When adequate reference points are available, the eye is very reliable for orientation. In the absence of a defined horizon, however, pilots may mistakenly choose some line of reference other than the real horizon.

Equilibrium is a state of balance between opposing forces. The three semicircular canals in the nonauditory portion of the inner ear provide balance information to the brain. The canals are filled with fluid and are situated at right angles to each other. One portion of each canal is enlarged. A mound of sensory hair cells is located in this enlarged area. Movement or body rotation tends to move the fluid within the semicircular canal, causing the hair cells to be displaced. Impulses are then transmitted to the brain and are interpreted as motion about a known axis. The hairs which project into the fluid are extremely fine; they bend with the fluid's slightest movement. The information transmitted to the brain is based on which direction these hairs are displaced. Your brain then computes the direction of body rotation.

Since each canal lies in a different plane, the semicircular canals can report three-dimensional rotation. The system works fine for short turns, but when the turn continues for any length of time, the motion of the fluid catches up with the canal walls. The result is that the hairs are no longer bent and your brain receives a completely false impression that the turning has stopped. Even in a gentle turn for just a few seconds, it is impossible for your semicircular canals to detect that you are actually in a turn. You can demonstrate this to yourself by making a shallow 10° turn in any direction the next time you fly.

Vertigo

Try to keep disorientation to a minimum when flying in poor visibility. Make shallow turns. Avoid moving your head or tilting your head more than necessary. Avoid looking up or down.

If you rotate rapidly and then suddenly stop, the fluid in the canals will continue to move. The hairs will continue to bend giving you the sensation of still being in motion. The movement of the fluid creates such a strong impulse that your eyes are affected. When the turn stops, your eyes continue to make short "sweeps" as if you are rotating. For example, we all can remember when, as children, we dashed out onto the playground and began turning around and around in circles until we made ourselves dizzy—and then laughed because our eyes wouldn't stop moving! We were experiencing true *vertigo*. Gentle turns and rapid turns to a complete stop can create similar illusionary impulses in the brain. Be alert for these illusions and *believe your instruments*.

The Leans

One type of illusion associated with balance and equilibrium is called "*the leans*." It's caused by correcting an abrupt change in an aircraft's bank or attitude slowly. After the correction is made, the pilot is left with the impression that the plane is not in the desired attitude. A pilot with "the leans" may feel compelled to roll the aircraft back into its original dangerous attitude. Pilots have been known to pull the wings off an aircraft trying to respond to this sensation. The illusion can be easily demonstrated on night flights, especially during inclement weather, such as thunderstorms. Most instrument instructors insist that their students experience this sensation during at least one of their lessons.

The tension of your various skeletal muscles also assists you in determining your position with respect to a point of reference. These clues can be misleading, however. Muscle senses may give you the impression of flying straight and level when in fact you are actually in a very well executed coordinated turn.

Spatial disorientation and vertigo are problems that usually show up under poor visibility conditions. Whenever you can't verify your orientation with your eyes, your body's equilibrium system is unreliable. Some pilots find it difficult to put their faith in dials and indicators, but during periods of poor visibility, the instruments in an aircraft are actually much more reliable than "human" equipment.

The instruments that help dispel illusions include:

⚠ Artificial Horizon

⚠ Turn Coordinator

⚠ Airspeed Indicator

⚠ Vertical Speed Indicator

⚠ Altimeter

Student instrument pilots are sometimes convinced that these instruments are lying to them. Don't fall into this death trap by thinking that your instruments are wrong. The incidence of actual instrument failure is virtually negligible.

Illusions

The following illusions are generally associated with flying. While some are rare, others are typically encountered frequently by pilots on mountain flights.

Spatial disorientation is an illusion of motion and position. It is most likely to happen when mountain terrain is obscured by poor visibility (e.g., false horizons caused by cloudy conditions), and can be prevented only by visual reference to reliable fixed points on the ground or to flight instruments.

Coriolis illusion is the mistaken feeling of rotating or moving about an entirely different axis. It can usually be prevented by avoiding sudden or extreme head movements, especially during prolonged constant-rate turns under IFR conditions or marginal visibility.

Graveyard spin is an illusion of spinning in the opposite direction. Generally it will occur when a proper spin recovery interferes with the body's motion-sensing system. A disoriented pilot will want to return the aircraft to its original spin. It can usually be prevented by close monitoring of aircraft instruments following recovery.

Graveyard spiral, an illusion which gives the pilot a false feeling of being in a descent with wings level, generally occurs during a constant-rate turn when the fluids aren't bending the hair cells in the inner ear. You can prevent it by close attention to aircraft instruments or outside reference points.

Somatogravic illusion, most common in high-performance aircraft, is a feeling of having a nose-up attitude during a rapid acceleration and takeoff. If not attended to, this illusion can result in a nose-down attitude. It can be overcome by giving proper attention to the airspeed indicator and artificial horizon.

Inversion illusion is the sensation of tumbling backwards following an abrupt transition from climb to straight-and-level flight. Pilots tend to push the plane's nose sharply downward, thus aggravating the illusion even further. Closely monitoring the artificial horizon and altimeter will generally fend off this illusion.

Elevator illusion may be caused either by gradually encountering an area of sink or an area of lift, or by suddenly experiencing strong updrafts and downdrafts—both of which are a frequent menace of mountainous terrain. The illusion gives the pilot the false impression of being in a climb or dive attitude. It can be prevented by monitoring the altimeter and the artificial horizon. In strong updrafts and downdrafts, the airspeed indicator is not particularly helpful in identifying elevator illusion because of the variation of the relative wind and subsequent ram-air effect on the pitot tube. The needle on the airspeed indicator will tend to fluctuate erratically. Likewise, the vertical speed indicator may also show fluctuations.

False horizon is induced by a combination of sloping cloud formations, an obscured horizon, a dark scene spread with ground lights and stars, and certain geometric patterns of ground lights. It results in the aircraft not being aligned with the true horizon. Pilots can combat this illusion by monitoring the artificial horizon and the directional gyro.

Autokinesis is the illusion in which stationary lights appear to move when you stare at them for several seconds. Following the imagined movement, you try to align the aircraft nose with the light. This illusion can be prevented by verifying your actual heading with the compass, heading indicator, and the stationary light. Avoid staring at just the one light. Pick out other lights to watch in your direction of flight or to the right or left of the aircraft nose.

Threshold deception is caused by atmospheric conditions and certain kinds of surface features seen by the pilot during landing. This illusion gives you the sensation of being too high, too low, too far, or too close to the runway threshold. Using a glide slope indicator or VASI (where available) and maintaining proficiency in landings will assist you in identifying this deception. Particularly in the mountains, you should overfly unfamiliar airports prior to final approach. Make a "low approach" and visually inspect the runway before landing.

Runway width illusion is common at remote airports with unusually narrow runways. The narrowness gives you a feeling of being too high on approach, so you mistakenly fly a lower-than-normal approach. Avoid this error by picking out and comparing objects on the ground and near the runway: trees, large rocks, signs, and—best of all—other aircraft or people.

Runway and terrain slopes can create the illusion of being too high or too low on approach. This illusion causes an inordinate number of crashes each year in the mountains. If you are uncertain of the runway slope, overfly it before landing. Pick a touchdown point in advance of landing. Plan your approach to that particular touchdown reference point rather than trying to analyze the whole runway during your approach to landing.

Featureless terrain, such as snow, water, large dark or white spots, or barren tundra, can give a pilot the illusion of being too high on approach. Similar to sloping runways, featureless terrain presents a serious danger to the unwary pilot. Practicing approaches at precisely 1.3 times the V_{so} of your aircraft will prove invaluable. Close attention to ground effect gives a pilot an awareness and control during descents to featureless terrain. Featureless terrain landings are discussed in greater depth in Chapter 6.

Atmospheric illusions. Rain on a windshield gives the illusion of greater height. Haze can cause the illusion of greater distance, and, during penetration, fog can present the illusion of pitching up.

Ground lighting, such as lights along a road or parking lot, can be mistaken for runway lighting. Lighted runways located within a vast open and dark area give you a feeling of less distance to the runway threshold. This is particularly true in desert regions. Overfly the field first, and set up a normal traffic pattern if terrain permits—this will really help a lot.

FLIGHT PROFICIENCY

Mountain flying, as mentioned earlier, is precision VFR flying; it requires a high degree of flight proficiency. Mountain pilots can and do experience a multitude of different situations, all in a relatively short period of time. They must master more than just the basic flight maneuvers. They must be completely familiar with their airplane and skilled in handling it. They must recognize and admit not only their own personal limitations, but also the aircraft's limitations in performing certain tasks. And they must be able to exercise good judgment quickly.

Proficiency implies that you have acquired a great deal of knowledge, possess an aptitude for performing certain tasks, and through training and practice have mastered specific techniques. Furthermore, it implies that you can demonstrate those skills and techniques with relatively the same degree of accuracy each time.

By way of an analogy, most of us would agree that athletes must constantly train for an event. Football players don't perform at their peak unless they practice, race car drivers don't make it to the winner's circle without preparing and drilling for the victory, and pilots are no different. Pilots must prepare constantly, training methodically, and practicing until maneuvers and procedures become second nature. It's not simply a matter of developing good habits. Habits can be performed without thought. In the mountains, flight procedures must be executed with a high degree of premeditation.

Proficiency naturally includes staying current on the approved flight maneuvers and procedures for each make and model you fly. Practice executing emergency procedures. Practice making different types of landings and takeoffs (short-field, soft-field, sloping runways, etc.). Know what the crosswind limitations are for your aircraft, and practice maneuvering in crosswinds—on the ground and in the air. Practice the different types of stalls; know how to recognize the special stall characteristics of the airplane you fly.

When the going does get tough, those pilots who are well trained and well "exercised" will perform best. They will be able to handle the situation and the airplane safely. The point can not be overemphasized—mountain flying requires a thorough knowledge of aircraft capabilities and precise manipulation of controls by the pilot.

In the final analysis, proficiency is a subjective term. Although there are lots of different tests which attempt to measure proficiency, only you know whether or not you are proficient.

Aerobatics and Soaring As Tools to Proficiency

Aerobatic training is an excellent way of learning about yourself and your aircraft. It helps fine-tune basic flight skills, and it builds self-assurance and courage. However, aerobatic experience is not necessary, and we are not encouraging you to perform stunts of any kind while flying in mountainous terrain.

Soaring can also prove invaluable in developing flight proficiency and safe mountain flying techniques. A glider rating will enhance your basic flight skills and knowledge, build confidence, and increase your overall understanding of aerodynamics. Unlike most power pilots, glider pilots see turbulence, mountain waves, and other natural forces as useful tools to flying.

is a lot like having an insurance policy. But, like STOL kits, horsepower only provides a margin of safety. The aircraft must still be flown and operated within specified limits.

A well-trained, proficient pilot can use the forces of nature to accomplish much the same thing expected from increased horsepower. In fact, pilots who routinely fly aircraft with less horsepower often find themselves in situations where they develop better pilot techniques. They know what they have to work with and are quick to use other, more natural forces to compensate for the lower horsepower.

"Horse Sense vs. Horsepower"

George Wastenott and Charlie Burnemgash planned a flight last summer to their favorite fish camp. The flight took them northwest of Anchorage a few hundred miles, but they had to cross the Alaska Range to get there. Wastenott owns

DHC-2 Beaver on wheels. This versatile aircraft has been used frequently in Canada and Alaska. It's a real heavy-hauler. (Steve Woerner)

and flies a 75-hp Taylorcraft and has a lot of mountain flying experience. Bur-nemgash, on the other hand, is a part-time pilot who regularly rents a 150-hp Super Cub and has very little experience in the mountains. They both departed Anchorage at the same time, en route to the fish camp located near McGrath.

Burnemgash decided to climb up and over the mountain range—the bigger engine made it possible—in a straight path to McGrath from Anchorage. Wastenott lacked the horsepower to climb up and over the top. However, since he knew how to read the mountains and its air currents he was able to beat Burnemgash to their destination. Wastenott flew along the windward side of the ridges, using the lift there to climb over the top of the ridges. By watching the signposts—clouds—he avoided time-consuming downdrafts. Burnemgash, with more horse-power, spent an inordinate amount of time fighting downdrafts.

The point, of course, is that good ol' horse sense outperforms horsepower any day. Extra horsepower is often nothing more than a good excuse for hiding poor piloting techniques.

LANDING GEAR

If you intend to land on hard-surface runways most of the time, tricycle gear will do just fine. However, if you land off-pavement a lot, a taildragger may be better suited to your needs. Taildraggers are capable of handling a more rug-ged landing surface than the tricycle gear. They are preferred by many experienced mountain pilots who must fly in and out of remote airstrips, often short gravel strips carved out of a river bed or forest. The height of a taildragger's nose helps keep the gravel out of the propeller. Also, many owners find that taildraggers are better adapted to skis. While skis can be fitted for tricycle-geared planes, they require some modifications.

Fixed gear seems to be preferred over retractable gear where landings must be made off-pavement and on rough surfaces. Fixed gear are able to withstand a bigger jolt and are readily adaptable to skis or floats. On snow-covered air-fields, the wheel wells on retractables often become ice-clogged, sometimes freez-ing the gear up inside the wells.

Wheel fairings (pants) on fixed-gear aircraft look nice, and most increase speed a little, but in the real world of gravel and mud strips, or during periods of ice and snow, the darn things can be a real liability. Besides clogging up with ice, snow, dirt, and grass, wheel pants make it difficult for the pilot to inspect the brakes—a vital preflight procedure when landing on short fields. Fairings limit tire size to stock conventional tires; owners are unable to install larger tires. If you plan to fly in the mountains frequently, take the fairings off during the wet and cold months, and put them into storage. Use them only seasonally. The trade-off here, honestly, is safety in exchange for a few extra knots of airspeed.

Cessna 206 Turbo Stationair on Wipline floats. A great combination. For land or lakes, the 206 can carry heavy loads and performs very nicely in the mountains. The fixed water rudder attached to the bottom of the tail provides extra stability. (Steve Woerner)

"Fiberfloats" channel the water under the float, creating lift rather than suction. The plane rides higher in the water on takeoff and lifts off at a slower speed. (Steve Woerner)

DHC-2 Beaver on 4580 amphibious floats. (Steve Woerner)

Floatplanes must be specially equipped with a "float kit." Among other modifications, a flange is welded to the main frame of the aircraft. (Steve Woerner)

Cessna on EDO floats. (Steve Woerner)

EDO floats. (Steve Woerner)

A fully hydraulic Federal wheel ski utilizing springs instead of bungee cords. (Steve Woerner)

Aircraft with conventional gear must also have a ski mounted on their tailwheel. This is a closeup of a Scott tailwheel with a wheel penetration ski.
(Steve Woerner)

A hydraulic ski can be repositioned beneath the wheel, in flight, for smoother landings on snow. (Steve Woerner)

A typical fixed-position wheel penetration ski. (Steve Woerner)

31

Hydraulic skis. Note how the wheels ride up on top of the skis for landings. (Steve Woerner)

TIRES

Large tires are great if you need them, but most pilots don't. Tire sizes like 650×6 or 850×6 are standard upgrades to conventional stock tires. Tires as large as 29 inches border on overkill in most cases. The standard upgrade to larger tires will offer more prop clearance on rough gravel strips, minimizing the amount of gravel thrown up into the propeller, even at low power settings.

Large tires do roll over debris and chuckholes better than smaller ones, but they also tend to grab more on asphalt runways. This is an important point for taildragger pilots; too much braking action will cause the airplane to nose over during wheel landings. For example, a Super Cub with big tundra tires that is equipped with extended-range fuel cells in the wings will have a tendency to nose forward during wheel landings if the nose is initially held at or near level.

There are a few trade-offs to mounting larger tires. Because they increase drag, they reduce cruise speed. Because of added weight, they reduce the useful load. And well, you might as well sell your wheel pants because they won't fit over big tires. Large tires cannot be mounted on retractable aircraft.

Piper Super Cub with tundra tires. (Steve Woerner)

In cases where large tires are called for, the best all-around rough-strip tire is the 850×6 (or 8) size. If they are not really necessary, keep your stock tires mounted.

HIGH WING vs. LOW WING

High-wing aircraft will give you better ground visibility in the mountains. This is not to say that low-wing aircraft are inadequate, but on soft or graveled strips their flaps and ailerons can be damaged from the mud, rocks, or slush thrown up by the wheels. This doesn't happen often though, and low-wing airplanes have some advantages, particularly if you like to camp in the mountains. Low wings make for a great camping table, and with a tarp thrown over the wing, they make a fairly good tent support. Of course, gas is much easier to check in a low-winged airplane, and the lower configuration gives better stability in windstorms (when tied down).

SHORT TAKEOFF AND LANDING (STOL) KITS

STOL kits are designed to increase takeoff and landing performance. They can be installed on most aircraft. Modifications include upgrading engines and adding one or more items to the aircraft structure. The most common change is made to the leading edge cuff and vortex gates located on top of the wing, and between the ailerons and flaps. These cuffs and gates can be added to the wingtips as well. This change is designed to keep the high-pressure air that flows under the wing from seeking the low pressure on top of the wing. The result is greater aileron effectiveness at slow speeds.

STOL kits should be viewed as nothing more than an added margin of safety between the stall speed of the aircraft and the approach or takeoff speed. It is true that the stall speed decreases, takeoff and landing distance decreases, and rate-of-climb capability is enhanced when STOL kits are installed. However, it does not follow that STOL kits allow a pilot to go in and out of extremely short fields. In fact, if a pilot must depend on the use of a STOL configuration to depart or land at a specific airfield, he is probably flirting with catastrophe.

STOL kits are not a requirement for mountain flying. They are generally expensive to install and are hard to justify when you consider the small difference in performance that is actually achieved. Evaluate them carefully before installing them on your aircraft.

Piper Super Cub on skis with a STOL kit installed. (Steve Woerner)

Cessna 170 with a STOL kit. (Steve Woerner)

PREFERENCES

Professionals lean towards using Cessna 180s, 185s, and 206s in the mountains, although the majority of these pilots will admit to having a Super Cub stashed somewhere. According to many pros, Super Cubs are great for checking out new landing sites for the heavier equipment. Bigger equipment like the 180s, 185s, and 206s are high-wing aircraft and have fixed gear, constant-speed propellers, and lots of horsepower. They are fairly expensive to operate but are easily adaptable to skis or floats. They give air taxi owner/pilots a payload of nearly 1,400 pounds and can be loaded to the brim with passengers and cargo.

Some folks prefer amphibians, like the Lake or the Grumman, or light twins. For mountain flying, the best plane for you is the plane that best suits your planned use. If you rent, it should be one that you fly often and are the most comfortable with. At a bare minimum, it should be one in which you can consistently demonstrate a high level of proficiency. Mountain flying can put you at your limits quickly.

35

Table 2-1. *Comparison of popular mountain aircraft.*
Data are approximate and not for use in flight planning.

Aircraft	Cruise Speed	Stall Speed	Takeoff Distance	Landing Distance	Useful Load
Super Cub	115 mph	43 mph	500 ft	885 ft	765 lbs
Citabria	112 mph	51 mph	890 ft	775 ft	670 lbs
Cessna 150	117 mph	48 mph	1385 ft	1075 ft	540 lbs
Cessna 180	162 mph	62 mph	1205 ft	1365 ft	1285 lbs
Cessna 185	169 mph	59 mph	1365 ft	1400 ft	1765 lbs
Cessna 206	163 mph	61 mph	1350 ft	1350 ft	1613 lbs

TABLE 2-1 offers a basic comparison of some of the most popular mountain planes.

INSTRUMENTS

Instrumentation can be broken down into flight instruments, engine instruments, and avionics. The primary flight instruments in any aircraft include an airspeed indicator, altimeter, turn coordinator, and compass. They are what enable the pilot to function in the three-dimensional environment of aviation. Flying in mountainous terrain requires only these basic instruments, but by adding other instruments you will have more information about your engine and more precise information about your location.

Instruments in the aircraft supplement—at times even replace—your human senses, especially in instrument weather conditions. More reliable than your own human equipment, they communicate the aircraft's attitude and movement about its three axes. What human sense can give the rate at which you are climbing or descending?

Except for the magnetic compass, the basic flight instruments are either *gyroscopic* or *pitot-static*. The gyroscopic instruments include the artificial horizon, directional gyro, and turn coordinator. They are driven by a jet of air or by electric motors. In air-driven instruments, the jet of air usually drives a vacuum pump which causes the gyro to spin, but sometimes the air jet is routed directly to the gyro. The pitot-static instruments operate by measuring the air pressure that surrounds an aneroid barometer, or *diaphragm*.

A broken or disconnected vacuum hose, or a deficient electrical supply, will severely retard or even disable the gyroscopic system. The pilot may notice a sluggish artificial horizon or a "wandering" heading indicator. A check of these instruments during taxi can avoid some unpleasantries after takeoff. Operating in and out of dirt or sandy airstrips, or smoking cigarettes in the plane, often lessens the life of the vacuum system. Check your suction gauge during preflight runup to make sure the system is working properly.

Pitot-static instruments depend on ram air (through the pitot tube) and/or aneroid barometers. They include the altimeter, vertical speed indicator (VSI), and airspeed indicator. A leaky aneroid wafer, an ice-clogged pitot tube, or an obstructed static port can be disabling to this system. Spiders and wasps are particularly fond of building homes in pitot tubes. Typically, pilot climbing or descending may know that the altitude of the aircraft is changing but the change doesn't register on the altimeter or vertical speed indicator. The airspeed indicator may respond with erratic deflections either up or down the scale. As added protection, some aircraft have backup systems: a switch-activated pitot heater which prevents ice from forming in the pitot tube and an alternate static pressure source, usually a valve under the instrument panel. If you fly an older aircraft and have a hammer on board, you can break the glass on the VSI—it's a low priority item.

Other instruments also serve as limited backups. For example, pitch information might be obtained by watching for increases or decreases in the airspeed indicator, the altimeter, and vertical speed indicator; bank information might be determined from the turn coordinator or magnetic compass. If you fly an airplane with a manifold pressure gauge, altitude information can be estimated.

The most revered instrument for mountain pilots is the airspeed indicator. This instrument gives the pilot a quick reference to areas of sink, lift, updrafts, downdrafts, and wind shear. Sudden changes in airspeed indications may occur in areas of extreme turbulence. Most importantly, whenever there is a change in the aircraft's attitude (without a compensating change in power) that change will register on the airspeed indicator.

NAVIGATIONAL EQUIPMENT

VOR, *ADF*, and *LORAN* are the systems used most often for general aviation navigation. There are some intriguing variations to these systems, and there are other more-sophisticated systems available to airlines and the military. But for general aviation, these three are the navigational basics. In the mountains, however, their use can be limited.

VHF Omni Directional Range (VOR)

VORs operate within the radio frequency range of 108.0 to 117.95 MHz. Most are equipped for voice transmission and all are equipped with Morse code identification. VORs are limited to line-of-sight reception. In other words, an obstruction between your aircraft and the VOR station will interfere with reception.

Obstructions include the curvature of the earth, and mountains. Reception also depends on your altitude and the class of station (see TABLE 2-2).

Table 2-2. *VOR reception.*

Navaid Class	Flight Altitude (AGL)	Reception Distance
T (*Terminal*)	1,000 - 12,000 ft	25 NM
L (*Low Altitude*)	1,000 - 18,000 ft	40 NM
H (*High Altitude*)	1,000 - 14,500 ft	40 NM
	14,500 - 60,000 ft	100 NM
	18,000 - 45,000 ft	130 NM

Military pilots use a navigational system called TACAN which is similar to VOR except that it utilizes signals in the ultrahigh frequency (UHF) band. When a TACAN facility is coupled with a VOR station, the facility is referred to as a VORTAC. Distance measuring equipment (DME), which is also line-of-sight equipment, operates only off of VORTACs or combined VOR/DME stations.

Area navigation (RNAV) systems do not require a track directly to or from a ground navigation aid. RNAV permits the pilot to navigate to displaced or "phantom" facilities commonly called *waypoints*. There are three RNAV systems presently in use: the course-line computer, Doppler radar, and inertial navigation systems. The latter two are not at all dependent on signals from ground-based stations.

Course-line computers, the type of RNAV found most frequently on general aviation aircraft, use signals from VORTAC stations. They must be coupled with DME equipment and are limited to line-of-sight reception from the actual ground-based facilities—not the phantom facilities.

Because of the limitations imposed by line-of-sight transmission, mountain fliers should only cautiously rely on VOR, DME, and RNAV for navigation.

Automatic Direction Finder (ADF)

One of the oldest types of radio navigation systems, ADF provides pilots with an excellent backup to their VHF equipment and can be used when line-of-sight transmissions become unreliable. An aircraft equipped with ADF can receive low and medium frequency signals (190-1605 kHz) from ground-based non-

VOR line-of-sight transmission.

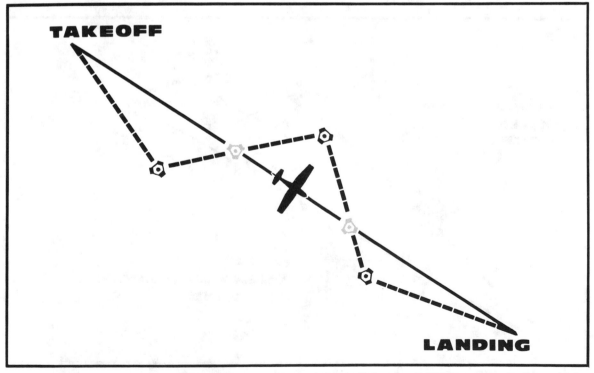

Course-line computers generate "phantom" VORs to permit direct flight.

39

directional beacons (NDBs), compass locators (stations co-located with ILS marker beacons), and commercial AM broadcast stations.

The needle on the ADF bearing indicator simply points to the radio station being received. The direction indicated is relative to the nose of the aircraft. Nearly all bearing indicators have an azimuth card. Some azimuth cards are fixed with 0° (or 360°) located at the top of the instrument. Others have a rotatable azimuth card (operated by a selector knob at the base of the instrument) for conveniently placing the aircraft's magnetic heading in the uppermost position of the instrument.

The *relative bearing* to the radio station is measured clockwise using 0° at the top of the instrument. For example, with the needle pointing at 270°, reading clockwise around the dial, the relative bearing is 270° from the nose of the aircraft.

The *magnetic bearing* to the radio station is the relative bearing plus the magnetic heading of the aircraft. For example, if the relative bearing to the station is 270° and the magnetic compass heading of the aircraft is 300°, then the magnetic bearing is 300° (270° + 30° = 300°). If the total is greater than 360°, subtract 360° from the total to obtain the magnetic bearing.

The relative bearing is how many degrees you have to turn your aircraft to align your nose with the radio station. The magnetic bearing is what your directional gyro or magnetic compass ought to read when you've completed the turn.

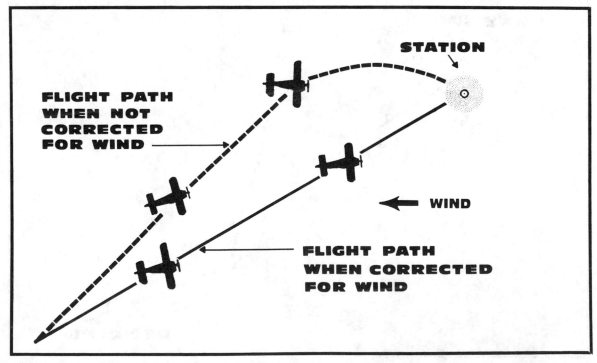

ADF ground track with and without wind correction.

People make too much of this. They say it is too complicated when really it is relatively simple.

The most common use for ADF is *homing* to the station; that is, if you keep the needle pointing to the uppermost position of the instrument—which represents the nose of the aircraft—you will eventually fly to the station. Unlike VORs, wind affects the actual ground track of the aircraft when you use ADF, so you should use the bracketing technique to correct for drift. However, even if you leave wind drift uncorrected you will eventually reach the radio beacon.

As with VORs, pilots can determine a position fix through triangulation with other NDBs, compass locators, or commercial broadcast stations. ADF is not affected by obstructions and is much more reliable than VORs in the mountains.

Although the benefits of ADF in mountainous terrain far outweigh its limitations, there are restrictions to its use. The *twilight effect* results when radio waves are reflected by the ionosphere and return to earth 30–60 miles from the station. This may cause the ADF bearing indicator needle to fluctuate erratically. The problem is most noticeable just before sunrise and just after sunset.

Radio waves will occasionally bounce off of sharply rising terrain (*terrain effect*). And magnetic deposits in some areas of the United States disturb reception and cause the needle to give a false reading. In the mountains, use only strong stations.

Thunderstorms with lightning may generate electrical disturbances in the equipment, causing the ADF needle to point to the lightning. Following the lightning flash, the needle will generally return to the direction of the station.

Coastline effect is the refraction (bending) of low-frequency radio waves over large bodies of water. ADF navigation is not recommended over oceans or very large lakes.

LORAN

An acronym for *LO*ng *RA*nge *N*avigation, LORAN is a system which measures the position of an aircraft (or boat) by computing the time difference between receipt of three different low-frequency radio signals. The radio signals are received from special *chained* stations operating in the 90–110 kHz frequency band, often hundreds of miles apart. Most ship pilots and many aviators are familiar with the term *LORAN-C*. The "C" is the modern designation for an earlier, less accurate version, called LORAN-A.

Within a chain, one station is designated as the *master* station and the other stations are designated as *secondary* stations. Signals transmitted from the secondary stations are synchronized with those transmitted from the master station. The time difference is computed by a receiver that compares a "zero crossing" of a specified radio frequency cycle within the pulses received from the master and secondary stations in a chain.

It all sounds pretty complicated but fear not; LORAN-C is "user-friendly." Preprogrammed at the factory or manually programmed by the pilot, it is very simple to operate. Behaving much the same as an RNAV system, it is a relatively inexpensive device that combines the benefits of using NDB-type radio transmissions with RNAV and DME capabilities, all for one price.

If you fly in mountainous terrain, LORAN-C is one of the best pieces of equipment you can have on board. Like ADF, it is not limited to line of sight. Like RNAV, which is coupled to VORs, it utilizes waypoints and phantom stations. And like DME, it measures distances. Its capabilities are built in to the unit and available to the pilot at the touch of a button.

LORAN-C is accurate to ¼ mile in primary areas, and some models have recently been approved by the FAA for en route IFR navigation. Instrument approaches with LORAN are still being tested.

The use of LORAN-C as a navigational tool is limited only by the areas presently covered by master and secondary "chained" stations. That is, certain areas in the United States and Canada currently do not have chains installed. These areas primarily include large sections of the Southwest and Plains states (called the "Mid-Continent Gap" by LORAN users), far northern and mid-continental Canada, and extreme northern Alaska (north of the Arctic Circle). Two new LORAN chains will be established, closing the "Gap" by the early 1990s.

COMMUNICATIONS EQUIPMENT

Communications equipment is an integral part of aviation today. Pilots should use it to their advantage. It is one of the many tools available to help ensure flight safety. As you fly through the mountains and canyons, you need to remember that communication, including radar, largely uses radio waves that are subject to the same limitations as VOR navigation equipment—namely, line-of-sight reception.

The standard two-way aviation radio continues to serve as the primary link between pilots and personnel on the ground. Transceivers in the VHF band (118.0 to 135.95 MHz) are limited to line of sight. Flying low in a valley will severely reduce if not totally block out all transmission and reception. Distance also is limiting, as the signals will not go beyond the horizon. Transoceanic airliners use high-frequency (HF) radios not subject to these limitations.

Transponders are airborne radar beacon receiver-transmitters. They receive radio signals from electronic interrogators on the ground and reply back to them with specific radio signals. Transponders, too, are limited to line of sight, so they do not communicate very well in the mountains.

Emergency locator transmitters (ELTs) are required to be on board most general aviation aircraft (see FAR 91.52). They are electronic, battery-operated transmitters which emit a DF (direction finding) homing signal. The signal is a distinctive downward-swept audio tone transmitted on 121.5 MHz and on 243.0 MHz. Transmissions are picked up by overhead satellites and then relayed to ground facilities. These facilities can be as much as several thousand miles away from the transmitter. Airline companies routinely monitor 121.5 and 243.0 MHz as well. ELT signals are transmitted in all directions, including upward, but are still restricted to line of sight.

SUPPLEMENTAL OXYGEN

Pilots planning trips into mountainous terrain may find it necessary to use supplemental oxygen. Sometimes the only way across a mountain range is up and over.

There are two types of supplemental oxygen systems: factory-installed fixed systems and the portable type. With fixed systems the cylinder is mounted in an aft or forward baggage compartment. The pressure regulator is mounted directly on the cylinder, and the flow is controlled by a knob usually mounted on the pilot's side of the instrument panel. Plug-in receptacles for masks may be positioned along the window sills or, for passengers sitting in the rear, in the headrest portion of the seat in front of them. Portable systems, on the other hand, are self-contained in a convenient carrying case. Each unit will have at least one cylinder, an oxygen supply gauge, an ON-OFF flow control knob, and two plug-in receptacles.

Oxygen systems are either *on-demand*, *constant-flow*, or *pressure-demand* systems. The on-demand type operates automatically each time a pilot or passenger inhales. The regulator valve opens automatically and allows oxygen to enter the lungs; the valve closes automatically when demand ceases. On-demand systems are more efficient—no oxygen is wasted—allowing pilots greater duration at altitude. Constant-flow systems supply a continuous flow of oxygen to pilot and passengers once the systems are turned on. Pressure-demand systems (usually found in turbine-class aircraft) increase the percentage of oxygen they supply as cabin altitude increases.

Aviation oxygen differs from medical oxygen and welding oxygen. The latter two contain excessive amounts of moisture which could freeze in the oxygen lines. Remember, as altitude increases, temperature decreases. A pilot may depart from a 5,000-foot elevation where it is 65°F, climb to 15,000 feet where it is 30°F, and have the oxygen lines freeze up. The higher moisture content in medical oxygen can be beneficial, however, on warmer days or in situations where there is no chance of freezing. It is less likely than aviation oxygen to dry out the sensitive membranes inside your nose.

A word of caution about the use of oxygen. When combined with petroleum-based substances, oxygen will severely irritate—even burn—skin. Make sure your face is clean. Be sure to remove lip balm, lipstick, petroleum jelly, or other related products before you put on an oxygen mask. Lastly, when combined with combustible materials, oxygen is highly explosive. *Do not allow smoking in the aircraft while oxygen is being used.*

3

Mountain Weather

WEATHER IS ONE OF THE MOST IMPORTANT ASPECTS OF MOUNTAIN FLYING. Pilots must be able to deal with a variety of weather conditions. Particularly in the mountains, weather systems tend to become localized and form smaller "mini-systems." These mini-systems can be completely different from and independent of larger, more regional weather patterns. It is not uncommon, for example, to find unstable air within the confines of a mountain range while the range is surrounded by stable air. Localized thunderstorms can form, and low-level wind shear may be present.

Mountain topography can and will create special wind conditions. Pilots approaching a mountain range may encounter mountain waves and other unique wind movements far from the range itself. And mountain passes often have accelerated winds caused by the venturi effect.

Airframe icing occurs frequently in mountainous areas. Pilots should know how to recognize potential icing conditions.

Pilots should adopt a constructive attitude towards weather. Aviation weather cannot be treated as purely an academic subject. Because pilots share the atmosphere with weather, one of the first determinations they must make is *weather to go or not to go.*

WEATHER SOURCES

Weather information comes from several sources: the National Weather Service (NWS), the Federal Aviation Administration (FAA), the military weather services, supplemental aviation weather reporting stations (SAWRSs), and pilot reports.

FAA Flight Service Stations (FSSs) are the primary source for preflight and inflight weather briefings. They provide three basic types of briefings:

Standard briefings automatically include information about adverse conditions, such as hazardous weather, runway closures, and inoperative navaids, plus SIGMETs, AIRMETs, and NOTAMs. They will advise pilots whether or not VFR flight is recommended and give a brief synopsis of the type, location, and movement of weather systems and air masses which may affect the proposed route of flight. Current conditions are provided, including SAs, PIREPs, and RAREPs. En route and destination forecasts are given, as are winds aloft, known ATC delays, and active military operations areas (MOAs).

Abbreviated briefings can be requested by the pilot who only needs an update on specific information obtained from an earlier standard briefing. Specific details are not provided unless the pilot requests a standard briefing or specific information about some aspect of the abbreviated briefing.

Outlook briefings can be requested when your proposed departure time is at least six hours away. This briefing should be used for planning only. Remember the old saying, "If you don't like the weather, then wait a minute." Weather does change and an outlook briefing can only give you an estimate of the future. It is only good at the time you receive the briefing, and its forecasts are very likely to change by the time you actually depart.

In flight or on the ground at an isolated airstrip, weather information may also be obtained from several types of continuous broadcast facilities or from Flight Watch or Flight Service. Continuous recorded broadcasts are made over selected VORs and low-frequency (L/MF) navigational aids (190-535 kHz). The low-frequency stations are less affected by mountainous terrain. If you're down in a valley, your best bet may be to tune in a L/MF weather broadcast station on your ADF receiver.

Another source of weather is Air Route Traffic Control "Centers" (ARTCCs). They are currently phasing in computer-generated digitized radar displays to replace older broadband radar displays. This new system provides controllers with two distinct levels of weather intensity. Although primarily for IFR flights, Center controllers will provide VFR pilots with in-flight weather advisories when requested. Keep in mind, of course, that their primary function is not weather reporting but aircraft separation. A safe route that will take you

Ed Gelvin clearing snow off his Citabria in his yard, central Alaska. (Galen Rowell / Mountain Light)

around significant weather areas may be requested from Center. They in turn offer pilots VFR flight following and vectors.

Other less professional sources of weather include state police (or sheriff's department), fish and game departments, the forest services, even local TV and radio stations. Here, in Alaska, weather reporting stations are sometimes scarce. More than once, we have found it necessary to telephone a local lodge owner in some remote location to ask what the weather is doing in his neck of the woods.

Pilots should understand that weather reports have limitations. The older the forecast, the greater the chance that some part of it is no longer valid or is simply inaccurate. It is prudent to regard weather forecasts as "professional advice" rather than absolute fact. According to the FAA, there is an 80% probability that poor weather will occur if it is forecast to occur within three to four hours of

WEATHER CHECKLIST

PREFLIGHT

☐ *Get a thorough weather briefing before departing.*

☐ *Give some thought to an alternate route should the weather at your destination turn out to be different from the forecast.*

☐ *Clean your windshield off so you can see the warning signs of bad weather.*

☐ *Give yourself some safety margin. In the mountains, expect conditions to be worse than reported or forecasted.*

☐ *Plan an alternative course of action in the event you encounter icing.*

☐ *Be cautious of a temperature/dew-point spread that is 5°F or less and shows indications of dropping further. It may be necessary to alter your course.*

EN ROUTE

☐ *Be prepared for severe turbulence, even icing, where frontal passage is forecasted.*

☐ *Near narrow passes and ridges, be alert for fast winds (the venturi effect), very strong downdrafts, and turbulence.*

☐ *Watch for ceilings to drop rapidly, as much as 2,000 feet per minute.*

☐ *Watch for low stratus in narrow canyons. These clouds reduce, or even block, the vision necessary for VFR flight.*

☐ *Know that the temperature/dew-point spread can close quickly to less than a 5°F spread, making fog very likely.*

the report. Common sense dictates that weather reports are going to be much more reliable in the first few hours of the forecast period than they will be as they get progressively older. This is especially true when distinct weather systems are present, like fronts, troughs, precipitation, and the like.

No matter how good a forecaster is, he cannot tell you when rain will begin to freeze, the location and occurrence of heavy icing, where severe turbulence exists, where ceilings will drop below 100 feet, or the onset of a thunderstorm before it begins to form. Visibility at the surface is also difficult to forecast—more difficult than forecasting ceiling heights. Forecasters can usually estimate ceilings to drop within ± 4 hours and can estimate their heights within ± 200 feet. Forecasters are able to guess snowfall within ± 5 hours of the actual occurrence.

During preflight planning, be sure to consider the big picture. Look at the trend of weather over a large area surrounding the planned route of flight. Weigh the available weather information and make an informed decision about your route of flight. If it turns out that the weather is indeed different from what you expected, you stand in good company. It is often different from what professional meteorologists anticipate. A prudent pilot will have an alternative flight plan in mind if there is the slightest chance the weather will turn bad.

PRESSURE SYSTEMS

Pressure systems can be a pilot's clue to what the weather will do and where it will move. There are five types of pressure systems:

- *LOW*—a center of pressure surrounded on all sides by high pressure (also called a *cyclone*)

- *HIGH*—a center of pressure surrounded on all sides by low pressure (also called an *anticyclone*)

- *TROUGH*—an elongated area of low pressure

- *RIDGE*—an elongated area of high pressure

- *COL*—the neutral area between two highs and two lows or the intersection of a trough and a ridge

Pressure patterns are identified on weather maps by lines of equal pressure, or *isobars*. Differences in pressure create what is referred to by meteorologists as a *pressure gradient force*. This force will move air from high-pressure areas to low-pressure areas. The closer isobars are on a weather map, the stronger the gradient force and the stronger the wind will be in that area.

The pressure in mountain passes, canyons, and valleys can be much lower than in surrounding areas. This lower pressure will be more pronounced closer to the surface. Wind will also be the strongest in these locations. The altimeter

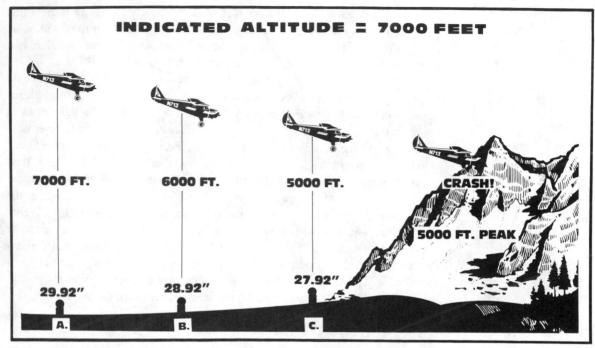

When flying from a High to a Low, look out below. (David Whitelaw)

in the aircraft will sense the change in pressure (from high to low) and will give an indication that the aircraft is higher than it actually is.

A change of one inch in pressure equals 1,000 feet of altitude (a likely occurrence in mountainous terrain). When flying to an area of lower pressure without resetting the altimeter for the pressure difference, the altimeter will read higher than you actually are.

See FAR 91.81 for altimeter setting requirements.

TEMPERATURE CHANGES

Temperature also affects altimetry. Air expands as it becomes hot and shrinks as it cools. As a general rule of thumb, for every 20°F or 11°C that sea-level temperature varies from the standard temperature (59°F), your altimeter reading will vary 4% from your actual altitude. (This should not be confused with the *standard lapse rate*, i.e., temperature decreases about 2°C with each 1000-foot altitude increase, on the average.)

The temperature in passes, canyons, and valleys typically will be colder than the surrounding air. An aircraft flying into this colder air will have an indicated altitude higher than its actual altitude.

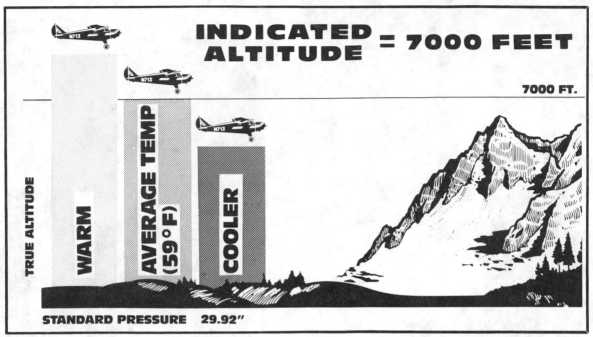

When flying from hot to cold, look out below. (David Whitelaw)

LOOKING AT CLOUDS

By watching clouds and other weather phenomenon, a pilot can predict the trend of approaching weather. The *sun dog*, a cirrostratus or luminous circle of ice crystals which appears to surround the sun or moon, is an early indication that weather in the mountains may change rapidly, even before any change in wind is noticed on higher peaks. In the Alaska Range, weather has been known to change from VFR to IFR conditions in less than 30 minutes after the appearance of a sun dog.

Clouds are more common indicators of what is happening in the atmosphere. They tend to form into the same patterns or types under similar atmospheric conditions, time after time. Consequently, clouds are defined by their appearance. They belong to two basic groups: the *cumulus* group, which billow, and the *stratus* group, which lay flat. The cumulus group is characterized by unstable, turbulent air, while the stratus group is usually characterized by stable, smooth air. Cumulus clouds produce large droplets of rain or hail, while stratus clouds produce small droplets of rain or snow. These groups are further subdivided by how high they extend into the atmosphere and by their characteristics.

51

A "sun dog" over the Alaska Range—a frequent sight in the mountains. It serves as a signpost for the wise pilot. (Galen Rowell / Mountain Light)

Height Classification

Cirrus clouds, and ones with a *cirro-* prefix, are very high altitude clouds. They usually form 16,500 to 45,000 feet above the earth's surface and are made up entirely of ice crystals. They give pilots advance notice of a warm front—as much as 24 hours before the front will pass. When they appear as frail tufts, fair weather will usually prevail. When they appear as *mare's tails* or dense bands, lower clouds and precipitation are probably approaching.

Clouds with the *alto-* prefix normally form between 6,500 to 23,000 feet above the earth's surface. They are composed mostly of water. This water may be supercooled—actually colder than 32 °F—and ready to freeze upon impact with your plane. *Altostratus* and *altocumulus* clouds which appear to be thickening usually mean that ceilings will lower and precipitation will begin within the next 6 to 10 hours.

A stratus or cumulus classification without the *alto-* or *cirro-* prefix describes clouds which form between the surface and about 6,500 feet and are made up of mostly water.

Mare's tails over the High Sierras near Big Pine, CA. (Galen Rowell / Mountain Light)

Characteristic Classification

A cloud type which has been prefixed with *nimbo-* or suffixed with *-nimbus* typically produces precipitation—rain or snow. For example, *nimbostratus* describes layered clouds which are producing rain. *Cumulonimbus* are billowing thunderstorms. Broken clouds are described by adding the prefix *fracto-* to the cloud type; for example, *fractocumulus*.

In the mountains, clouds can localize within the physical barriers. They can obscure ridges and peaks. Stratified layers can appear horizontal when in fact they are angular to the true horizon. Sunlight that penetrates through clouds is diffused, eliminating shadows and sorely reducing depth perception. Diffused light also causes dark colored objects, buildings, even people on the ground, to appear as though they are floating.

Cloud Heights

In the absence of a professional weather briefing, pilots can estimate the height of clouds above the ground with a fair degree of accuracy using the temperature/dew-point spread method. *Dew point* is the temperature at which,

Fitzroy, Patagonia. (Galen Rowell / Mountain Light)

when air is cooled, it becomes so saturated with moisture that the moisture becomes visible. When the spread is less than 5 °F, clouds (or fog) are likely to form. Because convective air cools at the rate of about 5.4 °F per 1,000 feet and the dew point decreases at a rate of 1.0 °F per 1,000 feet, it can be said that both are converging at a rate of about 4.4 °F (2.5 °) per 1,000 feet.

At an airport, for example, where the temperature is 47 °F and the dew point is reported to be 40 °F, the spread is 7 °F. By rounding the rate of convergence to 4 °F and then dividing that into the spread, the result will be the approximate height of the cloud base in 1,000s of feet: $7 \div 4 = 1.75 = 1,750$ feet. If the temperature and dew point are in centigrade, divide the spread by 2.2 to get cloud height in 1,000s of feet.

Temperature *inversions*, in which temperature increases with altitude, will give pilots a problem in using this method to calculate cloud height. Inversions can also cause strange things to happen. Particularly during the winter in arctic regions, they can cause a mirage to form—mountains look twice their normal size or seem to have flat tops. Or the inversion may make objects on the horizon appear as though they are above the horizon. Refraction of light as it passes through the inversion is responsible for these illusions.

Table 3-2. *Characteristics of Stratus and Cumulus Clouds.*

CHARACTERISTIC	STRATUS	CUMULUS
Stability; Turbulence	Stable; Smooth	Unstable; Rough
Precipitation	Continuous; Small Uniform Droplets	Showery; Large Droplets
Visibility	Poor	Good (Except in Areas of Precipitation)
Icing	Rime Ice; Snow	Clear Ice, Rime Ice, or Hail

CLOUD HEIGHT CLASSIFICATION

———————— 45,000 FEET ————————

CIRRUS · CIRROSTRATUS · CIRROCUMULUS

———————— 16,500 FEET ————————

ALTOSTRATUS · ALTOCUMULUS

———————— 6,500 FEET ————————

STRATUS · CUMULUS

———————— GROUND LEVEL ————————

CUMULONIMBUS

Overleaf: Mt. Cook, New Zealand. (Galen Rowell / Mountain Light)

Inversions—aloft and surface.

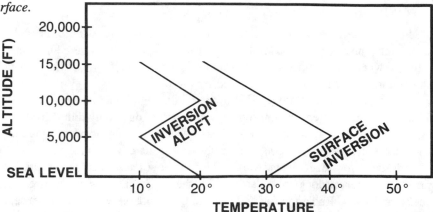

Icing

Airframe icing is a major aviation weather problem. It is difficult to forecast because its intensity can vary considerably under identical conditions. Accumulation rates may vary between one-half inch per hour to as much as one inch per minute. Icing can significantly reduce lift, substantially increase drag, plug up the pitot tube, clog the carburetor, and cause a whole host of other problems that can make flying difficult if not impossible.

Basically, for ice to form on an aircraft, visible water must be present and the temperature of the aircraft surface must be at or below freezing. Where temperatures are below 32 °F (0 °C) in clouds, icing is probable. Where temperatures are below 0 °F, the clouds themselves will have ice crystals.

Clear ice forms when large supercooled water droplets hit the aircraft when flying through rain or cumuliform clouds. Clear ice appears transparent or translucent and forms a glassy coating along the aircraft surfaces. It follows the shape of the airfoil and can eventually creep back to the ailerons, rudder, and elevator surfaces. It is hard, heavy, and sticks like glue. Even airborne/deicing equipment has trouble removing it from leading edges.

Rime ice, on the other hand, forms when small water droplets strike the aircraft when flying through light rain or stratiform clouds. Rime ice appears milky white, opaque, and granular. It is lighter in weight and more easily removed by airborne deicing equipment than is clear ice.

Clear and rime ice can form on aircraft surfaces as mixed ice. This usually occurs when droplets of varying sizes strike the aircraft, creating a "mushroom" shape along the leading edges.

High cirrus over the Grand Cathedral in the K-2 region of Pakistan.

(Galen Rowell / Mountain Light)

Fog

Fog, a surface-based cloud composed of either water droplets or ice crystals, is one of the most common and restrictive weather hazards encountered in flying. It forms when the temperature/dew-point spread is small, most often when temperatures are cool. Fog is classified by the way it forms.

Radiation fog is a relatively shallow fog. Although it can be dense enough to totally obscure the entire sky, sometimes radiation fog is only thin "ground" fog. This type of fog, which almost always occurs at night or near daybreak, is formed when the earth cools the air above it to its dew point.

Advection fog forms when warm moist air moves across colder ground or water. It is common along coastal areas and in the mountains along rivers, streams, creeks, and lakes. Radiation fog and advection fog look very much alike except that radiation fog is associated with land (only) and clear skies, while advection fog is associated with water courses and often accompanied by cloudy skies.

Upslope fog forms when moist stable air is swept upslope in the mountains and is cooled adiabatically as it rises. When the upslope wind stops, the fog usually dissipates. Often very dense, upslope fog can extend upward to rather high altitudes.

Precipitation-induced fog can form when warm rain or drizzle falls through cooler air. This type of fog can extend over wide areas, closing down airports for long durations.

Ice fog occurs in cold weather when the temperature is below freezing and water vapor changes directly to ice crystals. It is more prevalent in northern climates and is similar to radiation fog.

Fog of any type can restrict or totally obscure a pilot's visibility. It can make takeoffs and landings impossible and is found often in the mountains. Watch your temperature/dew-point spread; a spread of 5 °F or less makes perfect conditions for fog formation. Don't ignore the possibility that your landing site might be fogged-in if temperatures are right. Pay particular attention to runways which are near, on, or around water courses. And remember that fog is likely to exist longer when skies are heavily overcast.

AIR STABILITY AND TURBULENCE

Generally speaking, *stable* air is a mass that resists any upward or downward displacement. *Unstable* air is a mass that allows an upward or downward disturbance. Unstable air will grow into vertical or convective currents. Air may be totally stable or totally unstable but usually there is a delicate balance somewhere in between. A change in the ambient temperature lapse rate of an air mass may tip this balance. That is, surface heating or cooling aloft may cause the air to become more unstable, or surface cooling or warming aloft may create greater stability in an air mass. Air may be stable or unstable in layers. A stable layer may overlie and cap unstable air; or conversely, air near the surface may be stable while unstable layers exist above.

Turbulence is almost a certainty where unstable air crosses uneven terrain; usually it is the most turbulent on the leeward side. On the windward side, air is pushed up rapidly, forming *updrafts*, then spills over and down the leeward side, forming *downdrafts*. Often the speed of a downdraft will exceed the capabilities of your aircraft to maintain altitude or to climb.

Nature provides clues to the presence of unstable air and turbulence. Where sufficient moisture is present in the air, convective clouds will form on the windward side of a mountain and over the crest, warning of unstable air. The mountain as a barrier tends to reduce instability by mixing the air mass as it crosses.

In stable air, the mass moving up the windward side is relatively smooth and tends to flow in layers. Here, the mountain, as a barrier, sets up waves in these layers which extend downwind from the leeward side, sometimes as much as several hundred miles or even more. The crest of the wave itself will extend well above the highest mountain.

You can help other pilots by reporting any turbulence you encounter to the nearest FSS or ATC facility. TABLE 3-3 shows the criteria you should use in classifying the intensity of the turbulence you report.

Table 3-3. *Turbulence reporting criteria.*

VARIATION IN AIRSPEED	INTENSITY	AIRCRAFT REACTION	REACTION INSIDE AIRCRAFT	REPORTING TERM—DEFINITION	DERIVED GUST VELOCITY CRITERIA
5-15 KNOTS	LIGHT	Turbulence that momentarily causes slight, erratic changes in altitude and/or attitude (pitch, yaw, roll). Report as **Light Turbulence;*** -or-	Occupants may feel a slight strain against seat belts or shoulder straps. Unsecured objects may be displaced slightly. Food service may be conducted and little or no difficulty is encountered in walking.	Occasional-Less than 1/3 of the time. Intermittent-1/3 to 2/3. Continuous-More than 2/3.	5-20 fps
NONE	LIGHT CHOP	Turbulence that causes slight rapid and somewhat rhythmic bumpiness without appreciable changes in altitude or attitude. Report as **Light Chop.**			
15-20 KNOTS	MODERATE	Turbulence that is similar to Light Turbulence but of greater intensity. Changes in altitude and/or attitude occur but the aircraft remains in positive control at all times. It usually causes variations in indicated airspeed. Report as **Moderate Turbulence;*** -or-	Occupants feel definite strains against seat belts or shoulder straps. Unsecured objects are dislodged. Food service and walking are difficult.	Occasional-Less than 1/3 of the time. Intermittent-1/3 to 2/3. Continuous-More than 2/3.	20-35 fps
SLIGHT FLUCTUATION	MODERATE CHOP	Turbulence that is similar to Light Chop but of greater intensity. It causes rapid bumps or jolts without appreciable changes in aircraft altitude or attitude. Report as **Moderate Chop.**			
EXCESS OF 20 KNOTS	SEVERE	Turbulence that causes large, abrupt changes in altitude and/or attitude. It usually causes large variation in indicated airspeed. Aircraft may be momentarily out of control. Report as **Severe Turbulence.***	Occupants are forced violently against seat belts or shoulder straps. Unsecured objects are tossed about.	Occasional-Less than 1/3 of the time. Intermittent-1/3 to 2/3.	35-50 fps
RAPID, EXCESS OF 25 KNOTS	EXTREME	Turbulence in which the aircraft is violently tossed about and is practically impossible to control. It may cause structural damage. Report as **Extreme Turbulence.***			Over 50 fps

*High level turbulence (normally above 15,000 feet MSL) not associated with cumuliform cloudiness, including thunderstorms, should be as CAT (clear air turbulence) preceded by the appropriate intensity, or light or moderate chop.

AIR CIRCULATION

Wind passes over terrain features much like water flows over and around obstructions in a stream bed. That is, air masses rise and fall following the shape of the terrain. This shape, along with the heating of the slopes exposed to the sun, makes wind movement fairly predictable.

Flights conducted along a line parallel to a mountain range are likely to encounter steady updrafts or downdrafts. It is advisable to move away from the range, either upwind or downwind a few miles, to avoid strong or sudden updrafts and downdrafts. Rain shafts can cause strong downdrafts as well as violent wind shear. A rain shaft that falls from a cloud but does not reach the ground (called *virga*) is just as likely to have violent downdrafts and wind shear as one which extends all the way to the ground. Remember, light single-engine airplanes and many light twins are incapable of "outclimbing" strong downdrafts, even at maximum power settings.

Wind through mountain passes can also be violent. Like fluids, wind will accelerate as it is being forced through a small orifice. The result, of course, is that you may find yourself "zooming" through a pass faster than you anticipated—or crawling along like a turtle.

The movement of wind is seasonal. Fall and winter typically have stronger winds than spring and summer. In the summer months, strong winds are found at higher altitudes with the surface areas more thermal.

Wind will differ between daylight hours and night. Because some slopes are sunny and others are shaded, unequal heating patterns are created. Surface wind in the valley will move upslope during the day—from around noon until evening. In late evening, following a period of rapid cooling, the wind changes direction and moves downslope. Winds aloft over the tops of the mountains can come from any direction at any time of the day or night. You can make early morning and late afternoon flights closer to ridgeline level than flights during the heat of the day. In the summer months, start as early in the morning as possible to avoid the updrafts, downdrafts, turbulence, and winds that get stronger as the day progresses.

Valley Winds

Some winds are actually created by the mountains and valleys. As the air next to a slope is heated by the sun, it becomes considerably warmer than the air at the same altitude away from the slope. The colder air in the middle of the valley is more dense and sinks into the valley. When the cold air sinks, it pushes the warm, less dense air on the slope up and over the slope, creating a *valley updraft*.

Shadows, and the cooling which occurs after sunset, cause the air near the ope to cool faster than the air in the middle of the valley. The cooler, more

This view of Mt. Everest indicates wind direction for the pilot. (Galen Rowell / Mountain Light)

dense air along the slope then sinks into the valley causing a *valley downdraft* near and along the slope. Air that flows down gentle valleys is called *drainage wind*. Any downslope wind that is warmed by the warmer, less dense air it replaces is referred to as a *katabatic wind*. As the air becomes warmer and drier, it is funneled down the valley, and picks up speed. These winds can reach extreme speeds, sometimes as much as 90 knots. Some examples include:

⚠ The Chinook Winds—warm winds that blow down the eastern side of the Rocky Mountains often stretching out from the mountains 100-200 miles

⚠ The Santa Ana Winds—warm winds descending from the high Sierras down into the Santa Ana Valley of California

⚠ The Chugach Winds—warm winds which funnel down the Chugach Mountains in southcentral Alaska into the Anchorage Bowl area

⚠ Glacier Spill Winds—cold winds that spill over ridges and are funneled down into a valley that has been warmed by the sun

Katabatic winds are capable of causing severe damage to your aircraft, whether it is airborne or tied down. Prevent damage by heeding any forecasts of katabatic winds.

Obstructions to Wind Flow

When wind strikes an object, like a mountain, a rolling hill, or even a building, *mechanical turbulence* is created. The intensity of mechanical turbulence depends on the speed of the wind and the shape of the object it strikes.

A fast wind moving up and over a small berm or hill is less likely to be as turbulent as a fast wind moving up and over a ridgeline or peak.

Where wind flows across a valley or canyon at an angle or perpendicular to the mountain, the mechanical turbulence will be more severe at the crest or top of the canyon than it will be down inside the valley.

Wind will accelerate when it is forced into a narrow area, such as between two peaks or through a mountain pass. This acceleration is called the *venturi effect*. Mechanical turbulence may also be present in mountain passes, depending on the shape of the obstruction.

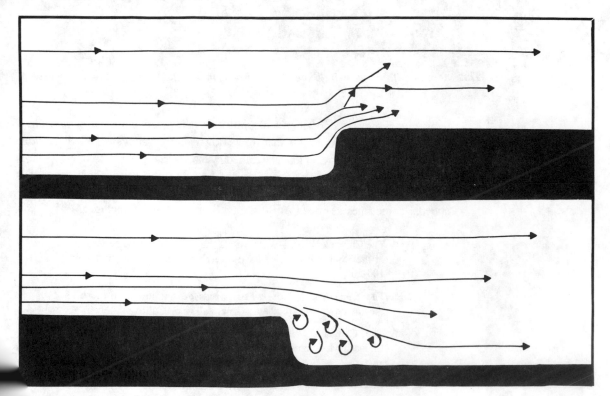

circulation over berms and small rises.

Wind blowing over the top of a canyon. The smoother ride is found low in the valley or well above the ridge lines, not at the same elevation as the peaks.

Wind blowing through a mountain pass is accelerated much like air through a venturi.

Stable, smooth air during climb-out is a good indication that turbulence over the mountains is at a minimum. Unstable, rough air during climb-out indicates the opposite. However, anytime you are flying over uneven terrain, hills, mountains, canyons, and the like, the chance for turbulence exists. Wind speeds and stability have a great deal to do with the frequency and severity of the turbulence you can expect.

In strong winds, begin your climb up the leeward side of a mountain well in advance of crossing—at least 100 miles before reaching the mountain if a mountain wave is present. Climb to an altitude of at least 2,000 (preferably 3,000–4,000) feet above the mountain tops before attempting to cross. Approach ridges at a 45° angle. This will give you an opportunity to retreat back to calmer air if it becomes necessary. *Do not* fly through mountain passes during periods of high winds; either climb above the pass or go around. Stay far enough away from the valley or canyon walls to avoid hitting the ground in the event that you get caught in a sudden downdraft. Always stay in a position to fly towards lower terrain.

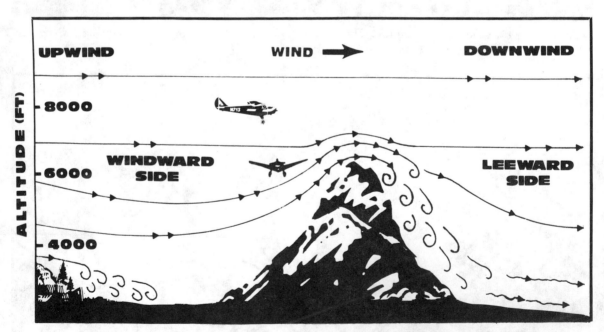

Turbulence can be avoided by flying 2,000 feet above peaks or along the upwind side of the ridge. Most turbulence is found on the leeward side of a mountain. (Adapted from David Whitelaw drawings.)

WINDS AND TEMPERATURE ALOFT

Wind direction, velocity, and temperature at specified altitudes are important tools for mountain pilots. They indicate whether conditions are right for mountain waves and tell a pilot how stable the air will be. Temperature inversions aloft indicate barriers of instability. Horizontal wind shear can be detected by studying winds aloft. For example, if winds at 6,000 feet are 360° at 30 knots and at 9,000 feet they are 180° at 20 knots, you can assume that horizontal wind shear will exist.

Special Mountain Weather: Wind Shear

Low-level wind shear occurs when the wind changes direction or speed over a very short distance. Under certain conditions, wind can change direction as much as 180° and speed more than 50 knots within 200 feet of the ground.

Some pilots mistakenly believe that wind only affects an aircraft's ground speed and drift. However, studies have shown that this is not true when the wind changes faster than the aircraft is capable of accelerating or decelerating, as is the case in low-level wind shear.

The most prominent causes of significant low-level wind shear are thunderstorms and certain frontal systems. Thunderstorms have complex wind systems. Typically, wind shear can be found on all sides and in the downdraft directly under thunderstorms. The actual wind-shift line or *gust front* can precede a thunderstorm by as much as 15 knots or more. Wind shear activity can be either vertical and/or horizontal to the surface. Pilots landing or taking off in close proximity to thunderstorms should do so with a great deal of caution. Low-level wind shear hazards probably exist. Be alert. Low-level wind shear caused by intense downdrafts called microbursts have been responsible for forcing aircraft as big as a DC-10 into the ground.

Weather fronts also cause high wind conditions. Warm fronts cause low-level wind shear just before frontal passage. Cold fronts cause wind shear just after passage. Warm fronts tend to have more severe wind shear than do cold fronts but the shear associated with cold fronts tends to linger longer in the area. One of the most difficult aspects of weather to forecast is precisely when the bad weather associated with fast-moving cold fronts and squall lines will begin. Typically, forecasters can say with some degree of reliability that fast-moving cold fronts and squall lines will pass a certain point within ±2 hours. This time frame can be forecast as much as 10 hours in advance of frontal passage. Meteorologists can forecast slow-moving cold fronts and warm fronts to pass a certain point within ±5 hours, and they can do this as much as 12 hours in advance of frontal passage.

Also stay alert for wind shear at airports near large bodies of water, like large lakes and oceans. The temperature difference between the land and the water can create high-velocity local airflows.

Low-level wind shear occurs often in mountainous terrain and is extremely dangerous. Mountain lee waves create wind shear at or near the surface and downwind of the wave (see ''Mountain Waves'' later in this chapter). Other indicators of wind shear include blowing dust, extreme variations in wind velocity and direction over short periods of time, surface temperatures in excess of 80°F, temperature/dew-point spread in excess of 40°F, airspeed fluctuations, and ground speed changes. Turbulence may or may not exist with wind shear conditions, but pilots should assume the worst where surface winds are strong and gusty under a frontal system.

Large airports have mechanical devices which assist in the early detection and warning of wind shear conditions, but mountain airports in remote areas will not have such devices. So you must carefully analyze weather charts for the possibility of wind shear if thunderstorms or frontal passage is expected in or around your flight path. And check for any wind shear PIREPS; most pilots are quick to report the presence of wind shear to Flight Service or ATC. If low-level wind shear is predicted, assume it exists.

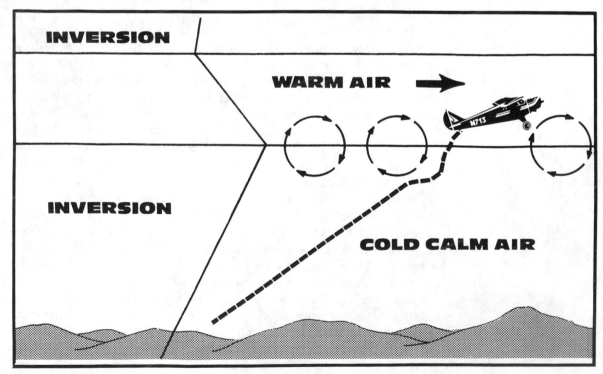

Wind shear can form in a zone between the calm air of an inversion and the strong wind above it. This condition is most common at night or in the early morning. It can cause abrupt turbulence encounters at low altitude.

THUNDERSTORMS

A developing thunderstorm has some very distinguishing features. The clouds look billowy and appear either white or dark grey in color, depending on the amount of moisture in them. At times they can even appear greenish—these are most often accompanied by hail, heavy rain, and moderate-to-severe turbulence. Pressure will drop rapidly with an approaching thunderstorm and then rise just as quickly with the onset of the first gusts, cold downdrafts, and rain showers. Thunderstorms frequently have lightning and low-level wind shear.

The Three Stages of Thunderstorms

A thunderstorm develops in three distinctive stages: the *cumulous* stage, the *mature* stage, and finally the *dissipating* stage. The cumulous stage is characterized by updrafts and is the "building" phase of a thunderstorm. As the updrafts begin, the air cools, causing condensation, i.e., clouds begin to form. Condensation releases latent heat which partially offsets cooling in the saturated updraft and increases buoyancy within the cloud. This increased buoyancy drives the updraft still faster, drawing more water vapor into the cloud and making the cloud self-sustaining. During the cumulous stage, clouds can grow at a rate of more than 3,000 feet per minute. As the cloud grows, water droplets build up in size. The cloud's weight above the freezing level increases and finally, ice forms.

The mature stage of a thunderstorm—the most hazardous stage—is marked by an anvil on the top of the cloud pointing in the direction of the storm's movement. As the water droplets become heavier, they begin to fall. Precipitation from the cloud base is a signal that downdrafts are developing and that the cell is in the mature stage. Cold rain in the downdraft retards compressional heating, and the downdrafts remain cooler than the surrounding air. The downward speed is accelerated, sometimes in excess of 2,500 feet per minute. The downrushing air spreads outward at the surface, producing strong gusty surface winds, a sharp temperature drop, and a rapid rise in pressure. The first gust is called a "plow wind," located at the storm's leading edge. Reciprocal updrafts and downdrafts follow, causing severe turbulence and vertical wind shear.

The dissipating stage is the final phase of a thunderstorm. Downdrafts associated with the mature stage eventually subside, and rain slows to a drizzle or stops completely. The cumulus cloud begins to break down, and the anvil becomes indistinguishable.

Thunderstorms are dangerous storms. Severe downdrafts can push your plane to the ground. Updrafts can carry you aloft before you can get out of the way.

Overleaf: A thunderstorm over the Palisades, John Muir Wilderness Area, California. (Galen Rowell / Mountain Light)

A thunderstorm on the Great Plains, Wyoming. (Galen Rowell / Mountain Light)

Tops can be anywhere from 10,000 feet to 65,000 feet. If you are carried up that high without oxygen, you're dead. The hail associated with thunderstorms can be the size of softballs. (Imagine the damage a hailstorm would inflict on an aircraft.) Clear and rime icing are possible, and low-level wind shear dominates the mature and dissipating stages.

In the mountains, thunderstorms should command more respect than at any other time. The close confines of passes, valleys, and canyons significantly reduces your ability to circumnavigate these storms. Steer well clear of thunderstorms.

THERMALS

The biggest difference between a glider pilot and a pilot who flies powered aircraft is that the glider pilot will hunt like crazy to find *thermals*, areas of convective lift. Thermals are caused by uneven heating of the earth's surface. They develop over any surface subject to uneven heating, such as an asphalt highway or a desert. Plowed fields, as another example, give off more heat than adjacent shrub- or tree-covered areas.

More thunderstorm activity over the Palisades. (Galen Rowell / Mountain Light)

The best time of the day for "thermaling" is the middle of the afternoon. Watch for blowing dust, smoke, or dust devils. These are good indicators. Look for birds; they are experts at finding thermals. Thunderstorms during the cumulous stage are easy targets for finding thermals—as long as you don't linger too long!

The angle of the sun can direct you to thermals. Since the sun rises in the east, early in the afternoon the eastern sides of mountains and hills, which have been heated longer, are best for obtaining lift. Later in the afternoon the southwestern slopes are best. In the mountains, low-power aircraft can use thermals to their advantage.

Thermals look like ice cream cones. They rotate either clockwise or counterclockwise. As you approach the heat source, watch to see which wing moves up first. Bank the aircraft in the direction of the wing that tips up. Always turn against the rotation; this allows you to maintain a slower ground speed while maximizing the lift.

MOUNTAIN WAVES

A *mountain wave* (or *lee wave*) is created when a stable air mass strikes the side of a mountain (range) and is forced up, over, and downwind. When the mass comes in contact with a layer of less stable air, a temperature inversion, or even the lower levels of the stratosphere, the layer of stable air will bounce until it comes in contact, again, with another lower layer of unstable air, temperature inversion, another mountain (range), or even the earth's surface. This bouncing or wave-like action may continue up and down for several hundred miles downwind of the first wave.

Mountain waves can be compared to the reaction of water to obstacles in a river bed. The current of water moves along steadily until it strikes a submerged log or boulder. The water is then forced up and over the log, falling down the other side until it bounces up off the river bottom. If we could "see" entire mountain waves, they would look very similar. For a mountain wave to form, winds of 15 knots or stronger must be blowing within 30° of perpendicular to the mountain range. Consult temperature aloft forecasts and reports. A constant temperature lapse rate indicates stable air and the possibility of mountain waves.

Mountain wave in Patagonia. (Galen Rowell / Mountain Light)

Wave Intensity

A mountain wave's intensity is described in terms of its *amplitude* and *wavelength*. The bottom of the wave is called the trough; the top is called the crest. Length of a wave is measured either between troughs or crests and is governed by the wind speed and stability of the air mass. A single wave can be anywhere from 2–35 miles long. The entire mountain wave system can normally extend downwind as much as 300 miles.

The amplitude of a lee wave is one-half the vertical distance between a wave's crest and its trough, but can be "guesstimated" to be the same as the distance between the mountain peak and the crest of the wave. Amplitude is governed by the size and shape of the mountain range or ridge, as well as by the wind speed and stability of the air. A more shallow layer of stable air and a more moderate wind speed will produce a greater amplitude than will a deeper layer of less stable air accompanied by strong winds.

Lee waves that constantly offer pilots the best lift are those waves with greater amplitude and shorter wave lengths. Although each wave differs, the average wave extends upward to 25,000 feet above the surface, but some have been known to reach altitudes of 46,000 feet.

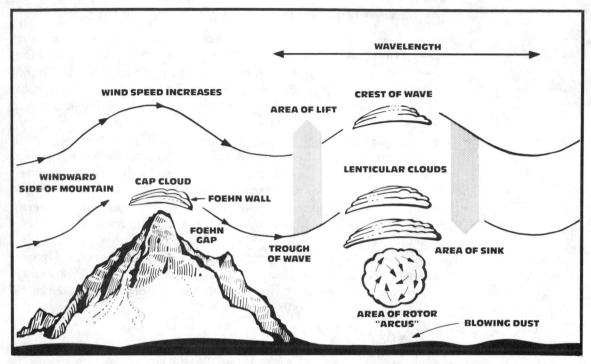

Mountain wave. Note the location of clouds. They serve as signposts for the pilot. (David Whitelaw)

Lee waves are strongest in the winter when stable air is more prevalent. Also the jet stream extends lower in the atmosphere during the winter. The stable air and the lower jet stream provide greater amplitude and shorter wavelengths, but, be that as it may, mountain waves can form during any season.

Associated Clouds and Signs

Stable air which is forced up a mountainside decreases in temperature. Where moisture is present in the air, clouds may form when the air mass reaches its saturation point (dew point). These clouds appear on both the windward and the leeward side of a range or peak. A *cloud cap* may form at the top of the peak and may even engulf the whole mountain summit. Sometimes these cloud caps look billowy like a cumulus cloud; other times they take on a smooth shape similar to a contact lens. Generally, the more stable the air flow, the smoother the cloud shape.

Just beyond the mountain on the leeward side, the air may be extremely turbulent. Here, eddies are formed. They can be identified by short cloud wisps that seem to first appear then disappear very quickly. These wisps are called *foehn clouds* and are formed by the compression of warm air as it flows down the lee-

A cloud cap over K-2 in Pakistan. (Galen Rowell / Mountain Light)

ward side of a mountain. That area between the leeward side of the mountain and the windward side of the first lee wave is referred to as the Foehn Gap. In the Foehn Gap, clouds can form and dissipate quickly or can completely cover the area in a very short period of time. Pilots who fly into the Foehn Gap should maintain an altitude equal to or greater than the highest peak to remain safely above this area. This is particularly important when flying parallel to a range on the leeward side. Flights conducted below the peak and within close proximity to the surface can expect moderate to severe (even extreme) turbulence.

"Lennies"

The *altocumulus standing lenticular* (ACSL) cloud, or "lennie" as it is called by glider pilots, is formed at the crest of the wave when the proper temperature and moisture conditions exist. Often it is possible to actually see the cloud form as the layer of air flows up and over the crest of the wave only to dissipate as it moves downwind in the wave. Lennies appear to flow, although they are really stationary clouds, hence the term "standing lenticular." Sometimes ACSL clouds

Overleaf: Lenticular clouds, Owens Valley, California. (Galen Powell/Mountain Light)

appear piled on top of one another, providing a visual display of the stable air sandwiched between the various layers of less stable air.

Avoid flying underneath "lennies." Always assume severe-to-extreme turbulence exists under standing lenticular clouds.

Rotor Clouds

The most severe turbulence a pilot will experience in a mountain wave can be found in and around rotor clouds.

Rotor clouds are warning signs to the mountain-wise pilot and are best described as turning, churning, bundles of severe-to-extreme turbulence. They may appear at elevations below the mountain peak and/or at the summit level. They are located below standing lenticular clouds. Frequently, they appear as harmless puffs or cumulus cloud formations, sometimes fragmented or torn apart from the main portion of the rotor. They may extend down to near ground level.

Avoid areas near rotor clouds, and remember that, if the air is dry, lenticular and rotor clouds may not form in the mountain wave, but the turbulence will still be there—and be just as hazardous.

If your destination airport is situated beneath standing lenticular and rotor clouds, be sure to look for signs of blowing dust or sand near the runway. Wind associated with rotor clouds can develop into wind shear and can be extremely hazardous. Divert to an alternate, if necessary. Any noticeable change in wind direction or speed could be directly related to wind shear or areas of sink from the wave itself.

Flight Precautions

Flight planning should include a review of winds aloft. These winds are the best source for determining if the right ingredients are present for mountain waves along your route of flight. Where the temperature lapse rate is stable and strong winds are forecasted at elevations below the mountain range, conditions exist for a lee mountain wave to form. Ask for pilot reports in the area and check hourly sequence weather reports for ACSLs. Check for cold fronts and the estimated time of frontal passage.

Throughout your flight, pay attention to cloud formations such as cap clouds, lennies, and rotor clouds. If moderate-to-severe turbulence or areas of sink which seem greater than your aircraft's climb performance are encountered, you should change your route of flight. Also, maintain an awareness of wind direction and flow.

IN SUMMARY

During preflight planning, gather as much information as you can about mountain weather and wind conditions *before* you fly through them. Make a point to find out the forecasted wind speed and direction and the stability of the air along your route of flight. Ask about cloud and turbulence reports in the area. Study satellite reports over your home computer or at the flight service station on your field; they can indicate the presence of mountain waves. Winds over the mountains in excess of 30 knots mean a pilot should expect some turbulence. Winds in excess of 40 knots mean it will definitely be turbulent. While en route, look for nature's warning signals—clouds which indicate the type of turbulence you are about to encounter:

⚠ Stratified clouds indicate smooth, stable air but possibly poor visibility in mountain passes.

⚠ Standing lenticular clouds indicate mountain wave.

⚠ Rotor clouds not only indicate mountain wave but usually extreme turbulence as well.

⚠ Convective clouds indicate rough, unstable air but usually good visibility.

⚠ Cirrus clouds usually signal a change in the weather.

White-outs occur when it begins to snow while flying under stratiform clouds. In the mountains, a white-out can happen very fast when temperatures are right for snow. There are no warning signals; snow simply starts to fall and can increase in intensity so fast that it makes visibility nil in all directions.

Mountain IFR. There is no reason to fly into or near clouds in mountainous terrain, unless you are passing over a mountain range on an IFR flight plan. Be aware, however, that if you find yourself in the clouds, clear and/or rime ice will form if temperatures are below freezing. This book is intended for flights conducted under VFR conditions. There is no such thing as IFR flight *in* the mountains.

4

Aircraft Performance

PERFORMANCE IS AFFECTED BY A MULTITUDE OF FACTORS. FIRST, THE AIRCRAFT has a set of specific performance characteristics—it will do specific things under given sets of circumstances. Temperature, wind, and altitude also influence performance capabilities, and even a pilot's flying technique is a factor.

As a pilot, you must know what your aircraft is capable of doing. This is particularly true in the mountains where precision flying is an integral part of safety. You must know each limitation of your aircraft. You must understand weather conditions that can impair performance. And you must develop proficiency in executing precise maneuvers to avoid further degradation of your aircraft's abilities.

EFFECTS OF WEIGHT AND BALANCE

Almost every aspect of aircraft performance is affected by an airplane's weight and the distribution of that weight. When coupled with certain atmospheric conditions and limited pilot experience, weight and balance become extremely critical.

Excessive weight:

⚠ increases the speed necessary for takeoff.

⚠ increases the distance necessary for takeoff.

⚠ reduces the rate and angle of climb.

Lukla, Nepal—elevation 9,000 feet. (Galen Rowell / Mountain Light)

⚠ reduces the service ceiling of an aircraft.

⚠ reduces cruising speed and range.

⚠ reduces maneuverability.

⚠ increases stalling speed.

⚠ increases landing speed and roll-out.

Being able to maneuver at slow speeds and having plenty of reserve power is very desirable in the mountains. An aircraft that exceeds its certified weight limit or whose center of gravity (CG) is outside the envelope will be hard, if not impossible, to control when moderate or severe turbulence is encountered.

Maximum gross weight is the sum of the certified empty weight and the useful load of an aircraft. Gross weight should never be allowed to exceed the certified limits specified in the airplane's weight-and-balance documents. No two airplanes are exactly alike—limits vary. Weight-and-balance documents for the aircraft you fly are required by regulations to be stored on board.

Empty weight is certified by the manufacturer and revised as necessary by certified mechanics after modifications. It includes the weight of the airframe, engine, standard and optional equipment, unusable fuel, undrainable oil, and hydraulic fluid.

Useful load is the difference between the certified empty weight and maximum gross weight. It includes weight of the pilot, passengers, baggage, usable fuel, and drainable oil.

Center of gravity is the point from which an aircraft, if suspended, would be in balance. An aircraft may be flown safely if its weight and center of gravity fall within its center-of-gravity *envelope*. Check the operating manual for the airplane you fly. No two types of planes are alike; each has its own unique center-of-gravity envelope.

Excess weight or an out-of-balance condition will impair performance and potentially affect the structural integrity of an airplane. Severe turbulence over mountain ridges and through mountain passes, as well as some special maneuvers mentioned later, may place the airplane at the uppermost limits of stress and performance. Be prepared. Don't overload your plane. Calculate the center of gravity, not only for takeoff, but also where it will be upon landing.

A nose-heavy aircraft, for example, requires additional airspeed to raise the nose off the ground and consequently requires additional runway length, perhaps more than the runway available. A nose-heavy aircraft is particularly difficult to handle on soft or muddy fields; keeping the nosewheel up and out of the mud may be a problem, increasing drag and decreasing performance. And even if the pilot is able to get the plane off the ground, it may not be able to achieve the climb angle needed to clear obstacles.

The stall speed of a nose-heavy aircraft is significantly increased. On landing, a higher approach speed will be necessary, which will increase the overall landing distance. If you fly a nose-heavy taildragger, you can easily nose over when you apply the brakes.

Landing a tricycle-gear aircraft on soft or muddy fields requires that you hold the nosewheel off the ground as long as possible to avoid damage to the nosewheel and strut. Excessive weight forward of the CG makes this difficult if not impossible.

Tail-heavy aircraft are equally as dangerous and are more common because the baggage compartment is usually located in the aft section. Although it is advantageous to load an aircraft so that the CG is in the aft portion of the envelope, make sure that the CG remains inside that envelope.

With a tail-heavy airplane you may find it impossible to hold the nose down on the horizon after takeoff, increasing the chance of a stall. During the initial rotation you may not have enough elevator movement to lower the nose and stay in ground effect at a slow speed. Flying out of ground effect without sufficient airspeed will result in a stall or "mushing" back to terra firma.

Never allow others to load your aircraft—not even other pilots. Although it is an easy point to overlook, make doubly sure that all passengers and all cargo are securely tied down. Expect turbulence over rough terrain. Improper loading or loose loads can cause the aircraft to become suddenly out of balance in flight, even though it may have been balanced properly on takeoff.

Heavy items, like bags of grain, concrete materials, or iron, should be located in the cabin so that the weight is in close proximity to the wing root. Do not place heavy items in the nose or aft baggage compartments. These compartments are best used for lightweight items like clothing, sleeping bags, tents, and the like. The best system is to follow the manufacturer's recommended placement.

Fuel management is a critical aspect of weight and balance. Some pilots fill their fuel tanks after each trip to avoid condensation buildup. This is particularly true in the "Sunbelt" region of the United States. In more northern reaches, where condensation is not such a problem, pilots tend to leave tanks half full so that fuel can be calculated and added after the main payload. If the fuel tanks are full and the gross weight exceeds the certified limits of the airplane, some of the payload must be taken off. If you need all of the fuel for your flight, you must remove baggage and/or passengers.

When loading hazardous materials, refer to *Code of Federal Regulations, Title 49*, for loading regulations and labeling instructions. Extra avgas is probably the most hazardous material the average pilot will ever load. When possible, use metal cans, not plastic. Plastic can produce static electricity which can cause an explosion. Pouring fuel into wing tanks or fuel cells with metal cans still carries with it a risk of static spark but to a much lesser degree. Use small containers for added safety.

External loading can be done safely, but it requires some added attention and special FAA approval for each operation. Plywood and canoes, especially, act like horizontal stabilizers or elevators when attached underneath the aircraft or to the spreader bars of floats. They may even block out certain portions of the elevator surface, making the elevator control less responsive. If you have a secret hideaway and intend to fly lumber or other large flat objects out to your site, there are some tricks to loading cargo on the outside of your aircraft. With lumber, drill some holes all the way through the bundle and bolt the entire stack together. Eye bolts are preferable because they provide an end to tie the bundle down with. For better flight stability, the side of a float rig is the best location for plywood bundles and canoes. The left side allows easiest pilot inspection.

Opposite: The late Don Sheldon, world-famous Alaskan bush pilot, refuels with a goat-skin chamois and gas can. (Galen Rowell / Mountain Light)

Cessna 172 on floats. Float support brackets can be used as luggage racks for hauling cargo and other materials up to your remote hunting/fishing camp. Be careful how you attach and secure external loads to the aircraft. (Steve Woerner)

EFFECTS OF ATMOSPHERIC CONDITIONS

Published aircraft performance figures are based on how an airplane performs in *standard atmospheric conditions*: sea level pressure of 29.92 inches of mercury and average standard temperature of 59°F (15°C). Variations from these conditions affect air density and will adversely or beneficially alter the performance of the aircraft.

Types of Altitude

True altitude is your aircraft's exact altitude above mean sea level. Altimeters do not indicate true altitude unless they have been corrected for nonstandard barometric pressure and temperature.

Indicated altitude is the altitude, above mean sea level, shown on your altimeter after the altimeter has been set to the local altimeter setting, that is, corrected for nonstandard barometric pressure.

Corrected altitude is indicated altitude which has been corrected for nonstandard temperature.

Pressure altitude is the altitude, in the standard atmosphere, where *pressure* is the same as where you are. The altimeter will indicate pressure altitude when 29.92″ is set in the Kollsman window. For example, if on a given day at a sea-level airport the barometric pressure is 28.92″, the altimeter will show approximately 1,000 feet when set to 29.92″. The pressure altitude is 1,000 feet because the pressure that the altimeter currently feels at sea level would be up at 1,000 feet *above* sea level on a standard day.

Density altitude is the altitude, in the standard atmosphere, where air *density* is the same as where you are. As altitude, temperature, and humidity vary, the air density varies. When air density varies, aircraft performance varies. Air becomes thinner or less dense when altitude, temperature, and humidity increase. As air becomes less dense, density altitude is said to increase, and aircraft performance is hampered.

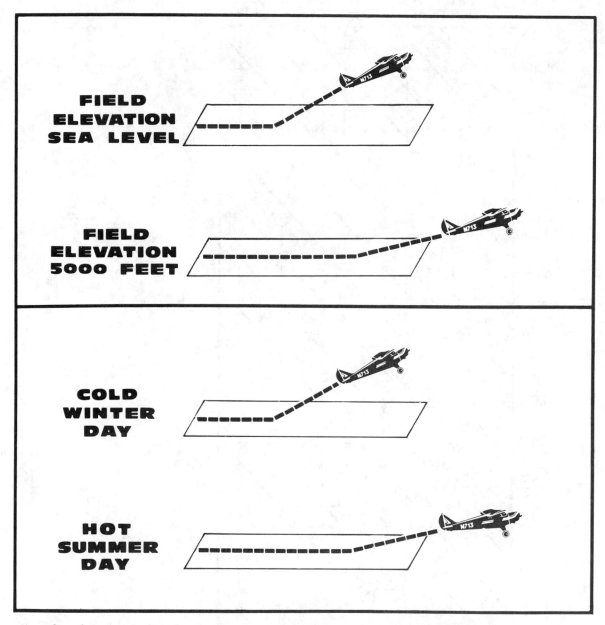

FIELD
ELEVATION
SEA LEVEL

FIELD
ELEVATION
5000 FEET

COLD
WINTER
DAY

HOT
SUMMER
DAY

Aircraft performance is directly related to the combined effects of altitude and temperature—density altitude. For example, a pilot attempting to take off from a 5,000-foot-high airport with a temperature of 80°F., would have a density altitude of 7,500 feet. The aircraft would perform as if it were at 7,500 feet! A typical ground run of 740 feet at sea level in cool air would become 1,300 feet at high altitude and temperature. Even if the pilot is able to depart the runway, it may be impossible to clear any trees, wires, fences, etc., at the end of the airstrip. It may also be difficult to clear any rising terrain in the departure path due to reduced climb performance.

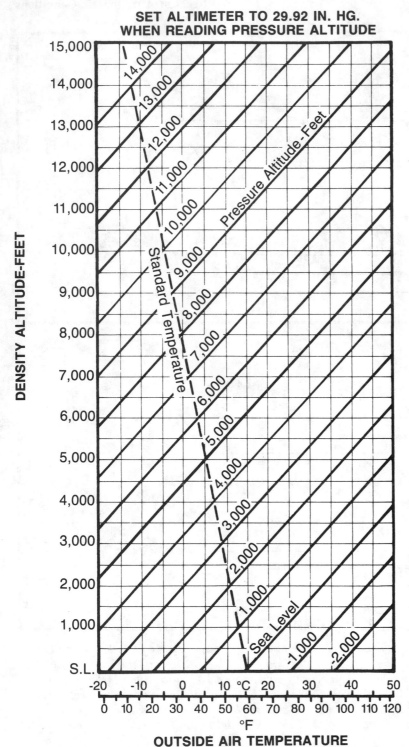

SET ALTIMETER TO 29.92 IN. HG. WHEN READING PRESSURE ALTITUDE

Density Altitude Chart.
(FAA)

DENSITY ALTITUDE-FEET

Pressure Altitude-Feet

Standard Temperature

Sea Level

°C

°F

OUTSIDE AIR TEMPERATURE

90

maneuvers and transitions are predicated upon how an aircraft will perform at these speeds.

Air becomes thicker or more dense when altitude, temperature, and humidity decrease. As air becomes more dense, density altitude is said to decrease, and aircraft performance is enhanced.

High density altitude (thin air) decreases available horsepower for normally aspirated engines and lessens propeller efficiency. If your airplane is turbocharged, high density altitude has less of an effect on performance, up to a specified altitude. An aircraft with a normally aspirated engine at a high altitude airport on a hot day will require more true airspeed for takeoff, and thus a greater takeoff distance. For example, an average small single-engine airplane requiring 1,000 feet for takeoff at sea level under standard atmospheric conditions may require as much as 2,000 feet for takeoff at a density altitude of 5,000 feet.

On the average, for each 1,000-foot increase in altitude, takeoff run will increase approximately 10% and true airspeed will increase about 2%. On one typical lightplane, for each 10°C (18°F) increase in temperature, takeoff run will increase 10%.

High density altitude also affects landing performance and lowers the service ceiling of an aircraft. Because it results in a significantly higher ground speed, pilots should be thinking of indicated airspeed when approaching a mountain airstrip and when otherwise operating close to mountainous terrain. Indicated airspeed tells a pilot how much relative wind is moving over the wings. Relative wind is necessary to produce lift. Paying too much attention to ground speed can be dangerous. When ground speed appears too fast, novice mountain pilots have a natural tendency to reduce airspeed so that they will see the ground moving at a speed normal for low-altitude airports. Unfortunately the result all too often is a stall.

Indicated airspeed is your best friend regardless of altitude or atmospheric conditions. But don't completely lose sight of the fact that, in high density altitude and/or no-wind conditions, you can overrun a short mountain airstrip. Moreover, you may not even be able to reach a high-altitude strip because your aircraft's service ceiling is much lower under high density altitude.

Close attention to factors like density and pressure altitude help ensure flight safety. Accurate altimetry is a pilot's link to the third dimension, not just with respect to takeoff and landing performance but to all aspects of mountain flight performance.

EFFECTS OF AIRSPEED

Aircraft performance is interlocked so tightly with airspeed that it is impossible to discuss one without the other. Pilots who elect to fly in the mountains must especially be aware of certain speed/aircraft configurations and the performance they render. It is important to *memorize* certain V-speeds. They must be ready to apply at the very instant you need them. The various mountain flying maneuvers and transitions are predicated upon how an aircraft will perform at these speeds.

Types of Airspeed

Indicated airspeed is the reading that a pilot gets directly from the airspeed indicator. It is uncorrected for variations in air density, installation error, or inherent instrument error.

Calibrated airspeed is indicated airspeed corrected for installation and instrument error. At some airspeeds, this error may be several miles per hour. The airspeed indicator should be calibrated periodically. Leaks may develop in the instrument, or moisture may collect in the tubing that runs from the pitot to the instrument housing. Dirt, ice, and snow near the entrance of the pitot tube may also interfere with accuracy. Vibrations may cause the diaphragm inside the instrument to become less sensitive.

True airspeed is calibrated airspeed corrected for variations from standard temperature and pressure—29.92 " and 59 °F. For a given true airspeed, indicated airspeed will decrease as altitude increases because air temperature and density decrease as altitude increases. Although flight computers are more accurate, a quick rule of thumb for determining true airspeed is to add 2 % to indicated airspeed for each 1,000-foot increase in altitude. True airspeed is used mainly for calculating time-and-distance problems in a cruise configuration (i.e., flight plans).

V-speeds are indicated airspeeds, although many are not marked on the airspeed indicator. They vary with each make and model of aircraft. The following are the most important V-speeds for single-engine aircraft:

V_a— The design maneuvering speed. It is the maximum speed at which you may use abrupt control travel without the likelihood of structural damage to the aircraft.

V_{fe}— The maximum speed allowable with flaps fully extended; the top of the white arc.

V_{le}— The maximum speed at which an aircraft with retractable landing gear may be flown with the gear extended.

V_{lo}— The maximum speed at which retractable landing gear may be operated (extended or retracted) in flight.

V_{ne}— The speed never to exceed in an aircraft. It is the red-line speed referenced on most airspeed indicators.

V_{no}— The maximum structural cruising speed. It is the speed that should not be exceeded except in smooth air and then only with caution; the top of the green arc.

V_{ref}— The meaning of this speed may vary. Often it refers to a final approach speed of 1.3 V_{so}. In this book it is defined as the minimum safe maneuvering speed in limited areas (see Chapter 5).

V_{so}— The minimum speed where a stall is imminent when flaps and landing gear are lowered in the landing configuration and the engine has been reduced to idle. It is the bottom end of the airspeed indicator (or the end of the white arc for aircraft equipped with flaps).

V_{s1}— As defined for the purposes of this book, the speed at which a stall is imminent when flaps and landing gear are not lowered; the bottom end of the green arc. Sometimes V_s is used synonymously.

V_x— The best-angle-of-climb speed; that is, the speed at which the greatest altitude gain is made for a given distance. It is the speed used to clear all obstacles.

V_y— The best-rate-of-climb speed; that is, the speed that produces the greatest altitude gain over a given period of time.

Table 4-1. *For flights in or around mountainous terrain, record and memorize these V-speeds for your aircraft.*

_____V_a—Maneuvering speed
_____V_{fe}—Maximum flaps-extended speed
_____V_{ne}—Never-exceed speed
_____V_{so}—Stall speed with flaps and gear down _____ 1.41 V_{so} *(60° bank)*
_____V_{s1}—Stall speed with flaps and gear up _____ 1.41 V_{s1} *(60° bank)*
_____V_x—Best-angle-of-climb speed *(sea level)*
_____V_x—Best-angle-of-climb speed *(10,000 feet MSL)*
_____V_y—Best-rate-of-climb speed *(sea level)*
_____V_y—Best-rate-of-climb speed *(10,000 feet MSL)*
_____1.3 V_{so}—Approach speed, level terrain
_____1.4 V_{so}—Approach speed, rising terrain
_____1.3 × 1.41 V_{s1} **(or 1.83 V_{s1})**—Minimum safe maneuvering speed in limited areas (see Chapter 5). Designated V_{ref} in this book.
_____Service ceiling *(ft MSL, density altitude)*

Best-angle-of-climb speed (V_x) will give you the most altitude for a given distance. If you think of the letter "A" in the word "Angle" as the mountain or obstacle that lies in your flight path, it might help you to remember that "best Angle" is the speed to get you over obstacles.

A speed slower or faster than V_x when V_x is called for is not recommended. With a slower speed, an aircraft is more likely to stall as drag increases when coming out of ground effect. With a faster speed, you cover more distance in less time, and consequently, altitude may be insufficient at the time you reach the obstacle. In some high-performance single-engine aircraft, best-angle-of-climb speed cannot be used for more than the time specified in the aircraft operating manual. Because the fuel tanks in the wing are below the line of gravity in these aircraft, fuel flow to the engine may be interrupted, causing the engine to starve for fuel and die.

Best-rate-of-climb speed (V_y) is the speed which will give you the most altitude over a given period of time. It is the speed to use after all obstacles are cleared and you are climbing up to cruising altitude. Best-rate-of-climb speed will keep the cylinders cooler.

Best-rate-of-climb speed and best-angle-of-climb speed vary with altitude, due to reduced air density. It is essential that you consult the aircraft operating manual (or call the manufacturer) to determine V_x and V_y for the specific density altitudes at which you will be flying. Don't expect to achieve maximum performance if you use sea-level numbers for your mountain flying.

The *service ceiling* of an aircraft is the density altitude at which an aircraft's best rate of climb is only 100 feet per minute. Check density altitude before you fly to determine if any of your planned altitudes exceed your aircraft's service ceiling.

Your aircraft's operating manual is the definitive source for service ceiling and V-speeds.

OTHER FACTORS

Aircraft performance is adversely affected by the equipment mounted on an airplane—the type of propeller, engine, the kind of landing gear, and so on. Floats, for example, can reduce speeds by as much as 30 MPH and reduce the useful load to the point where only one occupant can be carried—the pilot. Before flying into mountainous terrain, check the operating manual or owner's handbook to see how these special items affect the aircraft you will be flying.

5

Mountain Maneuvers and Operations

WITH TODAY'S SOPHISTICATED ELECTRONICS, FORGETTING THE FUNDAMEN-tals of flying is easy. The basics are important in all phases of flying but especially so in the art of mountain flying—one of the most demanding forms of VFR flying.

Mountain flying, sometimes called *drainage flying*, draws heavily from the old school of pilotage and dead reckoning. Flying skills and daily practice must be interwoven with modern pilot techniques. An awareness of aircraft maintenance and a thorough understanding of aircraft systems, speeds, weight and balance, mountain weather, and topography are musts for pilots. In addition to these ingredients, a mountain flier must possess a professional attitude towards flying, and above all, he or she must have common sense.

Practice. Being in good physical health is not enough. Nor is possession of ''adequate'' flying skills. As with all professional pilots, mountain fliers must develop an intuitive feel for the aircraft. They must sharpen their senses, particularly those of sight, sound, and touch. Through repetition, practice, and experience, pilots fine-tune their senses and skill level. Pilots who practice maneuvers often, and constantly analyze their own performance, will develop a certain ''feel'' for the airplane and be able to consistently demonstrate a high level of skill and performance. In the mountains, this is critical.

Develop Procedures Not Habits. Habits are nothing more than involuntary behavior. Dogs and cats develop habits. Procedures are a series of steps in a routine that should be consciously followed. When procedures become habits, that's bad.

Take the floatplane pilot, for example, who takes off from a hard-surfaced runway with retractable Wipline floats. Takeoff is made with wheels down. On landing at a nearby lake, he would be in trouble if he followed the standard *GUMPS* procedure. Habitually going through *G*as (on), *U*ndercarriage (wheels down), *M*ixture (full), *P*rop (forward), and *S*witches (on) can prove fatal. A wheels-down landing on a lake would probably flip the aircraft upside down. You must think about what you are doing.

Use and follow procedural checklists. Always use a *written* checklist—the one for the specific aircraft you are flying. Never rely on your memory. It is easy to forget items when you are tired, fatigued, or simply distracted. Maneuvering in the mountains can be both fatiguing and distracting.

Mountain flying requires pilots to make decisions about field elevation, crosswind component, runway slope, density altitude, even the length of the airstrip. If there is the slightest doubt that takeoff, flight, and landing cannot be done safely, then don't do it. Arguing with common-sense judgments could be disastrous. Tape a placard to your instrument panel which reads, *If in doubt, don't.*

TERRAIN FLYING

Terrain flying has been called contact flying, contour flying, drainage flying, or simply mountain flying. No matter how you may have heard it referred to, it is a specialized form of VFR flying and requires that you be able to recognize and identify unique landforms and key landmarks quickly. As a mountain pilot you must make judgment calls and react to situations miles ahead of the terrain.

Ridgelines and angles change rapidly as you fly from one valley to the next. The natural horizon will likely disappear when you fly up a canyon. So you must be able to recognize a *false horizon* and distinguish it from the natural horizon.

False horizons give a lot of pilots trouble. It's an easy mistake to make when flying up gradually rising terrain, when ridge lines change, when the slope and contour lines angle up or down, or when clouds obscure peaks and ridges. A natural reaction is to try to fly the aircraft level with the slope of the terrain—the false horizon. In gradually rising terrain, you may put yourself in a constant climb, and airspeed may bleed off, causing the aircraft to stall. The best cure is to maintain plenty of altitude when entering unfamiliar mountain passes or canyons. Always position the aircraft so that you can make a turn to lower ground. If you and your passengers are on a photographic flight, or you are simply inspecting a mountain airstrip, fly the aircraft in a downhill direction.

Use Your Senses

Your natural senses are your best friends in mountain country. Sight, sound, and feel are your personal links to *attitude* flying—flying without reference to the aircraft's instruments. Let the instruments serve as a backup to your senses when you're in rugged terrain. Power setting, airspeed, and attitude can all be identified to some degree by sight, sound, and feel.

Have you ever placed an aircraft in a slip or a skid? If so, you've heard the sound of air buffeting the fuselage. Pitching the aircraft up or down changes the noise level, while cruise flight has a monotonous humming sound. In addition to the changes in the sounds you hear, an uncoordinated aircraft will feel different. And, as airspeed bleeds off, the controls will feel more sluggish, less responsive.

A keen sense of sight will convey invaluable information—not only about dangers that may be coming around the next bend, but also information about the terrain itself. Very few routes *cross* mountain ranges. They go through the mountains by way of *passes*. Because most routes follow rivers or creeks, mountain pilots should be able to recognize the direction of flow in streams and rivers. Remembering that water flows downhill may give a confused pilot the edge needed to fly out of a valley.

In flatland, many rivers twist and turn with so little gradient that it's difficult to tell which way the water is running. In such cases, it's worth noting that floating trees that are snagged on sandbars nearly always have their root ends pointed upstream. It's also a good indication that the ground below you is relatively flat.

By watching the terrain behind an upcoming pass or ridge you can determine whether or not you are high enough to cross a peak or ridge with adequate clearance. If a landform beyond the pass tends to become smaller as you approach a pass, then you are too low. On the other hand, if the landform increases in size as you approach, the aircraft is high enough to adequately clear the obstacle (more on ridge crossings in a moment).

Sight also aids coordination. Remember during pilot training how your instructor pointed out wingtip movements? By watching the wingtip during the entry to a steep turn you could tell that the turn entry was coordinated when the wingtip moved straight down towards the ground object that formed the center of your turn. If the turn was developing into a skid, the wingtip seemed to slide backwards; or if the turn was developing into a slip, the wingtip moved forward. Keep your attention outside the cockpit. In the mountains, your instruments' visual attraction can easily become a fatal distraction.

Overleaf: McKenzie Mountains, Northwest Territories, Canada.
(Galen Rowell / Mountain Light)

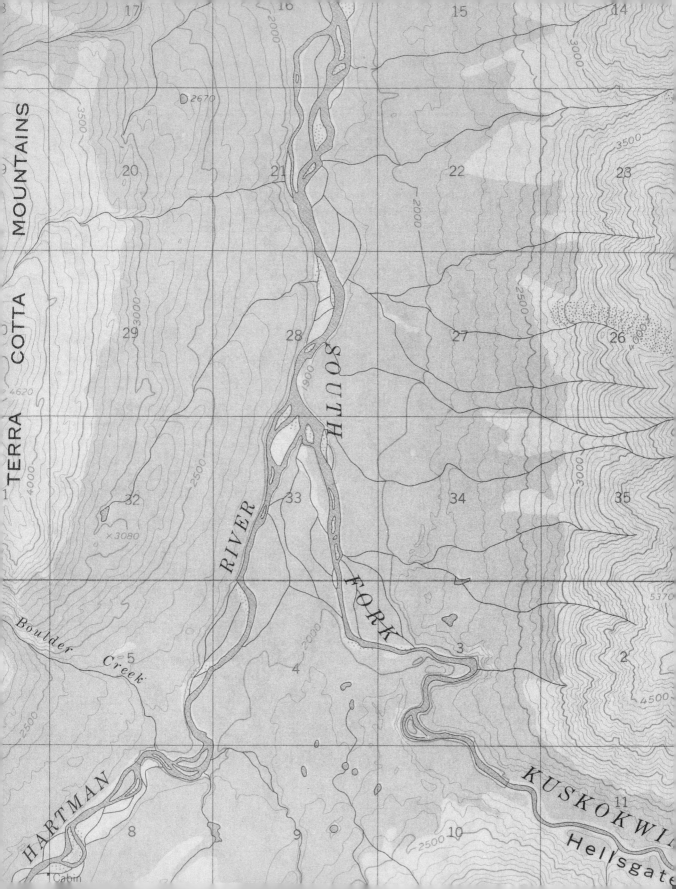

Charts and Maps

All pilots are trained to read and use aeronautical charts such as Sectionals and WACs, and we don't recommend abandoning them. But, there are other charts and maps which may be more useful in the mountains. Some of the best for terrain flying are U.S. Geological Survey (USGS) maps. These maps provide greater detail than Sectionals or WACs and are drawn a much larger scale. Contour lines, peak elevations, lake elevations, ridgelines, mountain passes, and distances between terrain features are more easily read off USGS maps. We strongly recommend their use as supplemental planning tools when terrain flying .

MOUNTAIN MANEUVERS

In this section we will present a few special maneuvers which will help the inexperienced mountain pilot build confidence. Understanding how the air moves through mountains, learning and understanding the fundamentals of flight, and most importantly, exercising good common sense are absolutely essential to learning and practicing these techniques.

Valleys

In valleys the wind moves upslope in the morning and downslope in the afternoon. Morning flights into a valley will accelerate you towards the valley's end more quickly; afternoon flights out of the valley will accelerate you towards the valley's mouth. Along the slopes, updrafts are more prevalent in the morning, and downdrafts are more prevalent in the afternoon. The severity of the wind currents inside a valley is governed largely by the overall weather conditions through which you are flying. If a front is moving through the area, expect updrafts and downdrafts to be severe and accompanied by moderate-to-heavy turbulence. Turbulence will be more noticeable within 2,000 feet vertically of peaks and ridges.

Depending on the physical terrain, pilots generally prefer to fly on the windward side of the ridge and slightly below the ridgeline. One side of the valley or the other affords a much better view of other traffic in the area and gives you the maximum amount of turning area should a 180° turn become necessary. Flying slightly below the peaks and ridges will generally give you the maximum altitude above the valley floor while placing the aircraft out of turbulence normally found near peaks and ridges.

Passes

In mountain passes, wind accelerates, causing a venturi effect much the same as air entering a carburetor. The accelerated wind may be as much as twice the

---

Mountain pilots use USGS maps to supplement their aeronautical charts.

intensity as the wind on either side of the pass. Anticipate the venturi effect, and plan ahead. Barometric pressure may also be affected, the altimeter reading will show an aircraft to be higher than it actually is above the Earth's surface. Adjusting the altimeter to *known* elevations of peaks and ridges or other landmarks can prove very helpful.

Never fly through a pass at a low altitude. Try to maintain at least 2,000 feet of clearance, and use your landing light to make yourself conspicuous to oncoming traffic.

Turbulence

Whether you fly in the mountains or out in the flatlands, at one time or another you have had to deal with turbulence. Although it seems more frightening when it happens in the mountains, it should not be a reason to avoid mountain flights. Encountering rough air does not mean that your airplane will come apart. Try to fly in the early morning, before clouds build and turbulence intensifies.

(David Whitelaw)

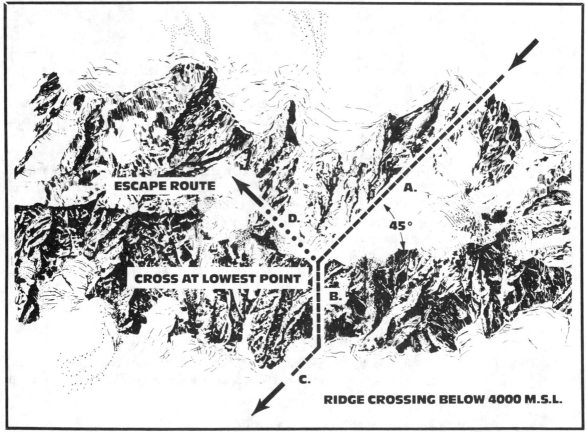

ESCAPE ROUTE

D.

45°

A.

CROSS AT LOWEST POINT

B.

C.

RIDGE CROSSING BELOW 4000 M.S.L.

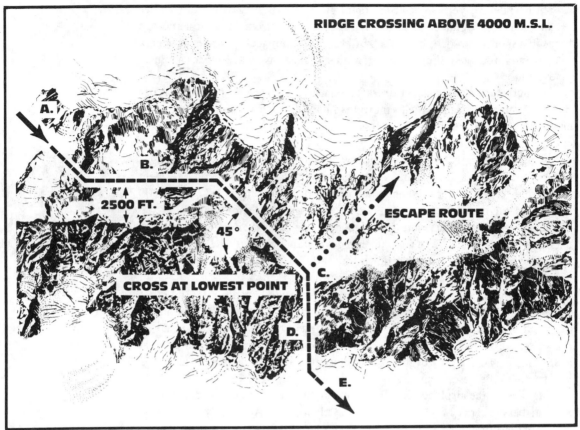

RIDGE CROSSING ABOVE 4000 M.S.L.

A.

B.

2500 FT.

45°

ESCAPE ROUTE

C.

CROSS AT LOWEST POINT

D.

E.

(David Whitelaw)

If you encounter severe or extreme turbulence, turn around or change your direction of flight to avoid that area. Turbulence almost always gets worse the further you go into the mountainous terrain. The following techniques apply to light-to-moderate turbulence, the most common type of turbulence.

Maneuvering Speed. For the mountain flier, *maneuvering speed* (V_a) should be one of the speeds committed to memory. It is the speed at which flight controls can be abruptly moved to full deflection without exceeding the maximum design load factor for the aircraft. It approximates 1.7 V_{so}, but please refer to the manufacturer's recommended maneuvering speed for the aircraft you fly. When you encounter turbulence, simply decelerate to maneuvering speed—or even a little slower. Then maintain pitch attitude as best you can with quick brisk control movements. If the turbulence increases in intensity, turn the aircraft with a shallow bank angle and fly out of the rough air. A flight through light-to-moderate turbulence may be a little rough but it is safe, as long as you maintain maneuvering speed or a little less.

In turbulence, vertical and horizontal gusts hit the wing at various angles causing the angle of attack to change abruptly and the aircraft to stall numerous times in a short time span, hence the buffeting or bumpy feeling you experience. In moderate-to-severe turbulence an aircraft may stall as many as 30 times a minute or even more.

Manufacturers calculate an aircraft's maneuvering speed based on its gross weight. If an aircraft that weighs more than its licensed maximum gross weight encounters turbulence, the situation could be deadly. In such a case, turbulence could tear the aircraft apart.

Ridge Crossings. Even the slightest wind will cause turbulence along ridges and create updrafts and downdrafts. Approach ridges at a minimum of 1,500 feet above ridge elevation for ridgelines that are less than 4,000 feet MSL, and 2,500 feet above ridge elevation for ridgelines that are greater than 4,000 feet MSL. Higher ridgelines are approached at a higher altitude because aircraft performance decreases with altitude. If you encounter a downdraft, keep the nose down. Don't stall. Use power to get out of the downdraft. At high altitudes, more power will be necessary to recover or sustain altitude.

Cross ridges at a 45° angle under normal light-wind conditions. And cross them after you have reached the necessary altitude—not while still climbing. If the wind is strong, parallel the ridges to test the strength of the wind and its associated turbulence. Then, if you judge it safe to cross, approach the ridge at a 45° angle. To maximize ground clearance, turn perpendicular to the ridge immediately after crossing.

Always fly the airplane in a position where turns towards descending terrain offer an easy escape. Maintain altitude, and plenty of it. Always fly on one side of a canyon or the other so that in the event a turn becomes necessary, you will have the maximum distance available to do so. In short, always leave the back door open and your stairway clear.

Peaks. When flying over the top of a mountain range, maintain an altitude of at least 2,000 feet AGL to help avoid moderate-to-severe turbulence.

Limited-Area Turns (Box Canyon Turns)

Most canyon-related accidents can be attributed directly to pilot inattention. Use checkpoints to keep you thinking ahead of the flight path rather than playing catch-up all of a sudden. Maintain plenty of altitude when flying over unfamiliar terrain. Be willing to turn around and go to the hangar if canyon weather begins to deteriorate.

Valleys tend to become narrower the farther you fly into them. Often, pilots find themselves "up the creek without a paddle," in a narrow box canyon without sufficient altitude to fly up and over the peak and without sufficient altitude to fly up and over the peak and without adequate distance between the canyon walls to execute a standard-rate 180° turn.

Table 5-1. *Radius of turn.* (A. White)

RADIUS OF TURN (FT)

AIRSPEED (KNOTS)	Bank Angle					
	20° (1.06 g)	30° (1.15 g)	45° (1.41 g)	60° (2.00 g)	70° (2.94 g)	75° (3.85 g)
40	394	245				
50	616	382	222	STALL (assuming 1-g stall speed of 35 knots)		
60	887	551	319	186		
70	1207	749	435	251	158	
80	1577	979	568	328	206	152
90	1996	1239	718	415	261	193
100	2464	1529	887	513	323	238
110	2981	1850	1073	620	390	288
120	3548	2202	1277	738	464	342
130	4164	2585	1499	866	545	402
140	4829	2997	1739	1005	632	466
150	5544	3440	1996	1154	726	535

NOTES:

1. Figures in parentheses are normal acceleration (g's) for turns at that bank angle in level flight.

2. In the cross-hatched area the stall speed is higher than the desired airspeed (assuming a 1-g stall speed of 35 knots). For example: at a 60° bank angle the stall speed is $1.41 \times 35 = 49.4$ knots (see TABLE 5-2); therefore if the pilot tried to fly a 60° bank in level flight at 40 knots he would stall. The cross-hatched area on this chart only applies for an airplane that has a 1-g stall speed of 35 knots. If the airplane you were flying had a 1-g stall speed of 55 knots, in a 60° bank the stall speed would increase to 77.6 knots (1.41×55); the minimum speed you should try a 60° banked turn is 80 knots.

3. Remember that the figures in this table are *radius* of turn. It will require twice these distances, laterally, to complete a 180° turn.

4. The figures in this table are mathematically correct for any airplane, except that the cross-hatched area has to be adjusted for each airplane. These numbers are correct in theory, but maybe not in practice. They do not compensate for a pilot's inability to hold a precise bank angle or precise speed, and those variations will affect the radius. Likewise, the numbers are also affected by wind, and by the pilot's technique in rolling into the turn; therefore, these numbers must only be used as a guide.

5. When making high-g turns, make sure that you have enough power to maintain the speed under the heavy acceleration; otherwise, you may inadvertently drift down to stall speed.

Hammerhead turns and chandelles are NOT the answer. Hammerhead turns are for airshows. In most situations a pilot will have neither sufficient speed nor all that much time to think about the hammerhead maneuver. If done wrong, a hammerhead turn will kill you. Chandelles are for practice. They are not the kind of maneuver you should practice in the mountains—or even in close proximity to the mountains, for that matter. What to do when you find out that you just turned up a box canyon and may have some difficulty getting out depends a lot on your altitude above the ground and the cloud ceiling in your area.

Whether you call it a *limited-area turn*, a *box canyon maneuver*, or a *180° escape*, this maneuver is one of the best for getting out of a tight spot. Safer than hammerheads and chandelles, the limited-area turn requires a turning radius of only one-third to one-half the turning radius of a normal 180° standard-rate turn, with a minimal amount of altitude loss.

Radius Of Turn. An aircraft's radius of turn is a function of bank angle and ground speed. A slower ground speed and/or a steeper bank angle will produce a smaller radius of turn. The aircraft's position is also important. As mentioned earlier, flights into any valley or canyon should be conducted closer to one side of the valley than the other, not down the middle of the valley. Preferably, flights into a valley should be on the downwind side of the valley. In the event a 180° turn becomes necessary, then a turn back, out of the canyon or valley and into the wind, is possible. In a limited area, a steep bank angle, a slow ground speed, and a turn back into the wind are desirable.

Radius is one-half of diameter. Looking at TABLE 5-1, a 130° turn at a speed of 90 knots in a 60° bank angle will have a diameter of just about 830 feet (2 × 415-ft radius). The shallower the bank angle, the greater the turning radius and, consequently, the greater the diameter of a 180° turn. The figures shown in parentheses are normal acceleration g's for turns in level flight. This particular table assumes a 1-g stall speed of 35 knots. The shaded area must be adjusted for each airplane.

The table is mathematically correct, but obviously no table can compensate for a pilot's inability to hold precise bank angle and precise airspeed, and those variations affect the radius. Actual turn radius will also be affected by wind and by the pilot's technique in rolling into the turn; therefore, TABLE 5-1 must only be used as a guide, and some tolerance must be applied in actual practice.

Minimum Safe Maneuvering Speed. As bank angle increases, so does the stall speed of the aircraft. In a 60° bank the stall speed is 41% higher than it is in level flight (see TABLE 5-2). In addition, as bank angle increases, so does the load factor of the aircraft. In a 60° bank the load factor is 2.00g, or twice the normal weight of an aircraft in level flight.

Because the stall speed and load factor increase with bank angle, it is important to have sufficient airspeed going into this maneuver. The best airspeed for a limited-area turn is what *we* call the *minimum safe maneuvering speed* (we

placard in your cockpit if you plan to fly in the

$$1.41 \ V_{s1} \times 1.3 = 1.83 \ V_{s1}$$

n, 1.41 V_{s1} equals the flaps-up stall speed in a 60°
1.3 equals V_{ref} or the minimum safe maneuvering
on the Wind-Slammer-200 is 49 knots IAS, by
1.41 we get a stall speed of 69 knots in a 60° bank.
all speed, 69 knots, times 1.3 we get V_{ref}, or about
fe maneuvering speed for the Wind-Slammer 200 in
be 90 knots.
reduced slightly if you can sacrifice altitude or if you
ion about flaps: with limited horsepower, high density
the aircraft is loaded at or near maximum gross weight,
d be dangerous.
soon as you realize that you've turned into a box can-
of-climb speed (V_x) to gain the maximum altitude.
s close as possible to the canyon wall, preferably the
anyon.

2. *Bank angle vs. load factor and stall speed.*

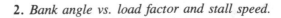

BANK ANGLE	AIRCRAFT LOAD FACTOR (g)	INCREASE IN STALL SPEED
0°	1.00	—
30°	1.15	7%
35°	1.22	11%
40°	1.31	15%
45°	1.41	19%
50°	1.56	25%
55°	1.74	32%
60°	2.00	41%
65°	2.37	54%
70°	2.92	71%
75°	3.86	97%
80°	5.76	140%

Once in position, assume V_{ref} for your aircraft, and roll into a 60° bank towards the upwind side of the canyon. Flaps may be used at pilot's discretion, depending on the aircraft's performance capabilities, density altitude, and weight. Partial flaps are generally safe to use for this maneuver.

The limited-area turn. (David Whitelaw)

designate it V_{ref}). Place it on a placard in your cockpit if you plan to fly in the mountains often:

$$V_{ref} = 1.41 \ V_{s1} \times 1.3 = 1.83 \ V_{s1}$$

By way of an explanation, 1.41 V_{s1} equals the flaps-up stall speed in a 60° bank. This stall speed times 1.3 equals V_{ref} or the minimum safe maneuvering speed. For example, if V_{s1} on the Wind-Slammer-200 is 49 knots IAS, by multiplying 49 knots times 1.41 we get a stall speed of 69 knots in a 60° bank. Then by multiplying this stall speed, 69 knots, times 1.3 we get V_{ref}, or about 90 knots. The minimum safe maneuvering speed for the Wind-Slammer 200 in a limited-area turn would be 90 knots.

This airspeed can be reduced slightly if you can sacrifice altitude or if you use flaps. A word of caution about flaps: with limited horsepower, high density altitude, or in cases where the aircraft is loaded at or near maximum gross weight, the use of full flaps could be dangerous.

The Procedure. As soon as you realize that you've turned into a box canyon, go to best-angle-of-climb speed (V_x) to gain the maximum altitude. Maneuver the aircraft as close as possible to the canyon wall, preferably the downwind side of the canyon.

Table 5-2. *Bank angle vs. load factor and stall speed.*

BANK ANGLE	AIRCRAFT LOAD FACTOR (g)	INCREASE IN STALL SPEED
0°	1.00	—
30°	1.15	7%
35°	1.22	11%
40°	1.31	15%
45°	1.41	19%
50°	1.56	25%
55°	1.74	32%
60°	2.00	41%
65°	2.37	54%
70°	2.92	71%
75°	3.86	97%
80°	5.76	140%

Once in position, assume V_{ref} for your aircraft, and roll into a 60° bank towards the upwind side of the canyon. Flaps may be used at pilot's discretion, depending on the aircraft's performance capabilities, density altitude, and weight. Partial flaps are generally safe to use for this maneuver.

The limited-area turn. (David Whitelaw)

If the canyon is extremely tight, it may be necessary to increase the bank angle beyond 60° and/or reduce the airspeed slightly. If you do this, altitude *must* be sacrificed to reduce the load on the wings and avoid stalling the aircraft. Allow the nose of the aircraft to drop below the horizon while, at the same time, increasing the degrees of flaps. The extra flaps will keep the airspeed from building too fast. Remember too, as long as the load of the aircraft is kept off the wings, the stall speed will not change.

After completing the 180° turn, add power, establish a positive climb attitude (transition to best-angle-of-climb speed), and retract flaps slowly.

Once you are proficient in this maneuver, it can be executed with less than 1,000 feet of altitude. However, if you have never done one and would like to practice, please practice at altitudes above 3,000 feet AGL.

Transitioning from Cruise to Mountain Landing

Landing on a mountain airstrip requires additional planning and attention to details. Executing a go-around may not be possible at the landing site you've selected. The airfield may be unimproved, short, soft, upsloped, or all of these. Mountain pilots must be prepared for the unexpected, particularly in remote areas. Moose, deer, or mountain goats may be grazing on the field you've chosen. Sound judgment and common sense are called for in these situations. And frequently, decisions must be made quickly.

Extra altitude is a real plus. You can see farther down the trail with lots of altitude. Not only can you find the airstrip easier, but with extra altitude, dangerous situations can be seen and avoided before they become a problem. Should an engine failure occur, extra altitude is like having added insurance, and generally, turbulence is less likely with altitude. Of course the question is, after you reach your destination, what do you do with all that altitude?

Three-Stage Transition. The *evaluation stage* begins when you arrive over the airfield. Generally it includes descending to a lower altitude, oftentimes by spiraling down over the top of the airstrip (positions A through C on the accompanying diagram). While circling the airfield into the wind and at a lower altitude, scrutinize the terrain features and the strip itself. Note whether the strip is an upslope landing site, a one-way-in box canyon site, and/or a short/soft field. Locate obstacles and areas of mechanical turbulence. Take special notice of any low-level wind shear activity. Measure density altitude and estimate wind direction and speed. Measure the runway length by timing the flight along a parallel track (position D). TABLE 5-3 provides estimated lengths based on ground speed and elapsed time. In poor lighting or near dusk it is particularly helpful to note the reciprocal heading while on this parallel track.

Positions E, F, and G are key points on the transition. Up to now you are not really committed to the landing. We call this the *escape stage*. It's the last segment where you still have the option to abort the landing.

Table 5-3. *Runway Length Estimation*

Ground Speed (Knots)	Distance (ft.) flown in:		
	30 sec	15 sec	10 sec
60	3040	1520	1010
70	3545	1770	1180
80	4050	2025	1350
90	4560	2280	1520

During the evaluation and the escape stages, take as much time as you need to determine the condition of the runway. Don't rush the approach, and don't feel compelled to make the landing if you see something that makes you uncomfortable. Be ready for the unexpected. It is not uncommon to spend as much as 20-30 minutes looking over an unfamiliar runway before committing to land on it. By the time you reach position H, your approach altitude, rate of descent, and airspeed should be set.

Runway at Gilgit, Pakistan. (Galen Rowell / Mountain Light)

The *commitment stage* truly begins after passing position H. In situations like the box canyon airstrip depicted, a go-around is not possible. After reaching this point, it is best to make the landing even when something unexpected happens.

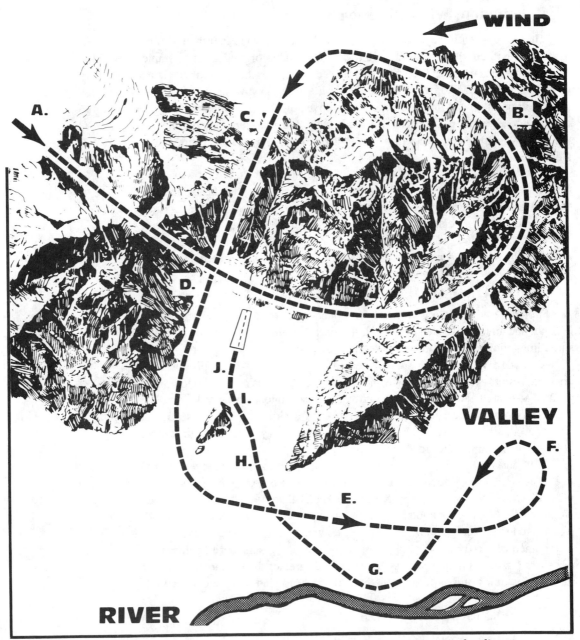

Transition to a mountain landing. (David Whitelaw)

Moose, deer, or other animals may wander out onto the runway, or at the moment of flare you may notice a chuckhole you missed during your overflight. Regardless of these misfortunes, it is safer to make the landing than to attempt a go-around.

Determining Wind Direction

There are several ways to determine which way the wind is blowing in the mountains without the benefit of a windsock. Over lakes or ponds, for example, the calmer end is upwind. Over heavily forested areas, the side of trees with shiny or glittering leaves is upwind, while the dark side is downwind. Fields of tall grass will point downwind, and smoke will move downwind, while animals, like cattle, moose, caribou, and deer will graze with their rear-ends pointed into the wind. Snow drifts and sastrugi (see Chapter 6) are created by winds. Their smooth sides face upwind.

For flights through mountain passes, valleys, and canyons, the snow blowing off the ridges or peaks helps pilots determine wind direction. Clouds also indicate wind direction. Mare's tails, for example, appear to wisp upward at their downwind end. If all else fails, you can find out which way the wind is blowing by performing a 360° standard-rate turn. Watch a fixed landmark and note the direction in which the plane is drifting. It's elementary but effective.

ANTICIPATING MOUNTAIN WIND SHEAR

Sudden wind shifts are common in mountainous terrain. When an aircraft in flight encounters a sudden headwind, the aircraft's nose will pitch up suddenly, sometimes violently, as lift increases.

A sudden tailwind—or sudden decrease in headwind—will produce the opposite results. Where low-level wind shear is present, it is not at all uncommon to have a strong headwind whip around very quickly into a strong tailwind. The result, of course, is a sudden loss of airspeed and altitude. Without sufficient airspeed, an aircraft may touch down short of the runway, threshold, or worse, crash into the famous 50-foot obstacle.

You must ensure that a sudden wind shift doesn't cause your airspeed to drop as low as your stall speed, especially at low altitudes. For example, if the power-off stall speed (V_{so}) for your aircraft is 45 knots, then normal approach speed would be $1.3 V_{so}$, or 59 knots. If you fly an approach into a 20-knot headwind, which suddenly turns into a 20-knot tailwind, your airspeed will drop instantaneously to 39 knots, and the plane will stall. To safeguard against this, use extra airspeed on approaches into headwinds greater than 5 knots. A simple method for doing this is to add one-half of the headwind component (HWC):

For flat terrain: $1.3 V_{so} + \frac{1}{2}$ HWC

For rising terrain: $1.4 V_{so} + \frac{1}{2}$ HWC

In the previous example, the recommended approach speed would be:

1.3 (45 knots) + ½ (20 knots) = 69 knots.

In a sudden 20-knot tailwind, airspeed would drop to 49 knots momentarily, but the aircraft would still be 4 knots above stall.

If you encounter wind shear on final approach and are unable to stay "ahead" of the aircraft, then abort the approach. Sometimes power, pitch, and airspeed change so quickly that even the best pilot will have difficulty keeping up with it all. Use common sense and pick another day to land at that site.

Airspeed should also be higher than normal when taking off into a strong headwind. By adding one-half of the headwind component to best-rate-of-climb (V_x) or best-angle-of-climb (V_y) speed, you can protect yourself against unexpected changes. If winds are strong and variable (if they vary beyond 90° to the runway), wait for more favorable conditions before taking off.

MOUNTAIN AIRFIELD OPERATIONS–IN GENERAL

In remote areas, takeoff and landing procedures, including the traffic pattern, must be modified to suit the airfield. High terrain, unusual wind currents, and airport layout will dictate how a pilot must adjust procedures.

When landing at unfamiliar or extremely remote airfields, we recommend that you overfly the field first. Have a look before you land. Any number of things could be wrong with the runway surface. On final approach do not allow terrain features to block visibility of the runway. Always keep the touchdown zone in view.

Do not follow other aircraft too closely; remember, most remote sites do not have taxiways, and landing aircraft will have to back-taxi on the active runway to exit. Circle the runway if the pilot in front of you needs more time to exit. Sometimes making S-turns will give the pilot in front of you enough time to exit, if you are already on final approach. The best rule of thumb at remote sites is to simply stay out of the area until the traffic has cleared the active runway.

Manipulate the flight controls smoothly. Use only enough control to get the job done. Any more than that will generally take away from valuable performance capabilities, especially during takeoff, landing, and maximum performance maneuvers. Rudder and aileron deflection increase drag. Drag increases takeoff distance and reduces overall performance.

Use brakes sparingly. This is especially important on snowy or icy runways or where standing water and slush are present. (Most mountain runways are not maintained, so it is always best to anticipate the worst.) Hard braking can melt snow and ice which will freeze again after takeoff, possibly causing the brakes to lock up on your return landing.

CROSSWIND OPERATIONS

Whenever possible it is best to take off into the wind, but more often than not, remote airstrips don't offer that luxury. In a crosswind, the objective is to take off or touch down without lateral drift, which would put excessive force on the landing gear and make ground looping of the aircraft a real possibility. Crosswind operations are accomplished by keeping the upwind wing down into the wind with the aileron, and by keeping the longitudinal axis pointed down the runway with the rudder.

Takeoffs

The crosswind takeoff requires that the ailerons be held into the wind and the takeoff path be held straight with the rudder. A tailwheel aircraft will require more downwind rudder since it will tend to weathervane into the wind while on the ground. Differential braking will be required to maintain runway heading until the rudder becomes effective.

In the case of strong crosswinds, the use of aileron control may cause the downwind wheel to lift off the runway first. In this case it may be necessary to complete the takeoff roll on one wheel. This should cause no unusual side load on the landing gear if you use the proper amount of aileron for correction of the existing crosswind and you hold the airplane on the ground until you attain adequate liftoff speed. This procedure will ensure that, when the airplane leaves the ground, it will remain airborne with the proper amount of drift correction established. This also will prevent side loads and possible damage that could be caused by settling back to the runway.

Climbout. As soon as the airplane is definitely airborne, make a slight turn toward the low wing, into the direction of the wind, to establish enough crab angle to take care of the drift. Then maintain a straight climb on this heading so that your ground track will be in line with the runway. This becomes extremely important when taking off from airports with parallel runways or from mountain airstrips surrounded by hazardous terrain—places where any unwanted drift could prove disastrous.

Landings

There are two well-accepted methods for crosswind landings. Both use the proper approach speed to landing which is 1.3 V_{so} plus one-half of the headwind component.

Using the first method, the upwind wing is lowered slightly, and the airplane is then slipped into the wind just enough to maintain the ground track in the desired direction (more on slips momentarily).

On landing, special attention is required to keep the airplane rolling in a straight line to prevent a possible ground loop. The aileron control should be

Go/No-Go Wind Component Chart

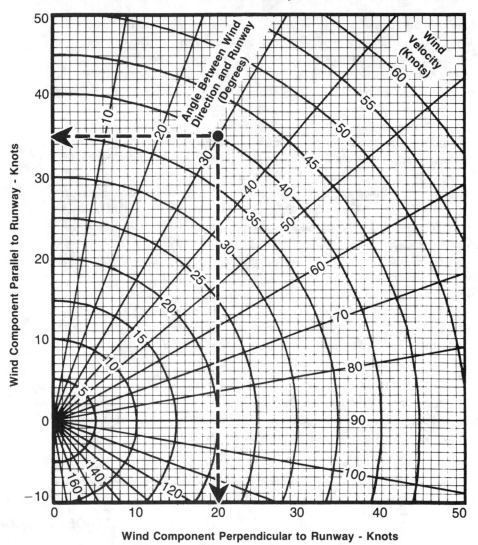

(1) *Determine the maximum 90° crosswind that you can handle (suggest 20% × stall speed). Place a dot on the 90° line at this value.*

(2) *Determine the maximum 45° crosswind that you can handle (suggest 30% × stall speed). Place a dot on the 45° line at this value.*

(3) *Determine the maximum headwind that you can handle (suggest 60% × stall speed). Place a dot on the 0° (vertical axis) line at this value.*

(4) *Connect the dots with a red line. Values to left of the red line are "go" wind velocities and directions. The dot shown is for a 40-knot wind with 30° angle.* (FAA)

held toward the upwind wing to prevent the wing from rising. It may be necessary to land on one wheel—the upwind one—if the wind is strong. One advantage of this method is that it leaves the upwind wing low at all times. This prevents a gust from upsetting the airplane close to the ground which could result in the downwind wheel or wingtip touching the ground first, a situation which could prove disastrous. This method is also advantageous for heavier aircraft whose responses are slow and reactions to sudden gusts are small.

With the second method, you maintain a heading (crab) into the wind so that the ground track is straight down the runway. Then, just prior to touchdown, you align the longitudinal axis of the airplane with the runway, using rudder control. If this is timed accurately, the airplane will touch down with its nose pointing exactly in the direction in which the plane is moving—straight down the runway. This method requires quick and accurate action to get the airplane lined up exactly at the instant of contact. Improperly executing this type of crosswind landing could impose severe side loads on the landing gear and result in a violent ground loop. The safety factor of a low upwind wing is absent in this method, and sudden gusts at the critical moment can more easily cause trouble.

With either method a slow airspeed will require more wind correction than a faster airspeed. Therefore, strong crosswinds require a higher approach speed, a higher touchdown speed, and more runway than normal.

Taxiing

In strong winds the pilot must *fly* the aircraft from tiedown to tiedown. Taxi no faster than you can walk, avoid using large amounts of throttle when turning, and position your flight controls as instructed in TABLE 5-4.

Table 5-4. *Control Positions During Taxi*

WIND DIRECTION	ELEVATOR POSITION		AILERON POSITION	
	TRI-GEAR	*CONVENTIONAL*	*TRI-GEAR*	*CONVENTIONAL*
Direct Headwind	Neutral	Up	Neutral	Neutral
Quartering Headwind	Neutral	Up	Upwind Aileron Up	Upwind Aileron Up
Direct Crosswind	Neutral	Up	Neutral	Neutral
Quartering Tailwind	Down	Down	Upwind Aileron Down	Upwind Aileron Down
Direct Tailwind	Down	Down	Neutral	Neutral
Note: In crosswinds, conventional-gear aircraft with large tail surfaces require substantial opposite-brake pressure to counteract weathervaning tendency.				

SLIPS

In the past, slips were regularly used to control landing descents to short fields. Now they are primarily used in crosswind landings and in emergency landings away from airports. The use of flaps in a slip should be minimal. Most aircraft are placarded against slips with full or partial flap settings because slipping with flaps destroys the airflow over the elevator.

When performing a slip, a pilot's attention should be kept on the centerline of the runway in order to detect drift. If the aircraft begins to drift, increase the amount of aileron. If the slip is too shallow, the drift will continue. The rudder is used to keep the longitudinal axis aligned with the runway centerline.

Slips to Landing

Since the purpose of a slip is to lower altitude without increasing speed, this maneuver is particularly important for aircraft not equipped with flaps. The performance of slips should form a regular part of your practice maneuvers for each type of aircraft you fly. There are several reasons for this.

First, if flaps should become inoperative, which is a rather common occurrence with electrical flaps as well as mechanical ones, the slip comes in handy for making short-field landings at high-altitude airstrips, or for forced landings.

Second, if you are on approach to a short-field high-altitude airstrip on a hot day, or at a high gross weight, the slip will increase your rate of descent just like flaps. If, however, you find yourself too low on approach you can recover from the slip without the additional loss of altitude you would suffer by taking off flaps. When you retract flaps from a full-flap position, you will lose additional altitude because of the decrease in the overall lift of the wing.

Third, if a forced landing is attempted because of loss of power, slipping the aircraft would be much more advantageous during descent to final until you can be sure that you can make the runway. After the runway is assured, you can use flaps.

Forward Slip. The *forward slip* is a valuable tool in mountain flying. This type of slip gives the pilot an excellent view of the landing area while providing a direct ground track to the runway threshold. Use of the forward slip is a safe way to get rid of unwanted altitude when you find yourself coming in too high, or when you want—or in the mountains, need—a shorter final approach.

A forward slip is achieved by lowering the left or right wing and deflecting the rudder in the opposite direction. In a slip, the elevator is used to control airspeed, while the power setting remains constant. To recover from a forward slip, relax rudder pedal pressure and level the wings.

Side Slip. An important crosswind tool, a *side slip* allows the aircraft's ground track to change without changing heading. Side slips are executed by lowering the wing in the direction of the wind and holding heading with opposite rudder. Side slips have little value in shortening distance or steepening glide path, but they make crosswind landings simple.

SHORT-FIELD OPERATIONS

In the mountains, short and soft fields tend to be the standard rather than the exception. The most common mistakes in making short-field takeoffs and landings usually involve poor planning, poor airspeed control, poor heading control, and poor power control. Practice is the best way to develop proficiency. Know your airplane. Memorize the appropriate V-speeds for short-field activity.

Short-Field Takeoffs

The primary object of a short-field takeoff is to gain sufficient liftoff speed in the shortest possible distance and clear all obstacles and terrain features that lie in your departure path. There are several different techniques. Some may be better than others. The best technique for you and the aircraft you fly will probably be the one recommended by the manufacturer.

Keep It Clean. An aerodynamically clean airplane will achieve best-angle-of-climb speed (V_x) in the shortest distance. Added lift may be gained through the use of flaps; if so, their use and setting will be recommended by the manufacturer. Maximum power should be applied as suggested by the manufacturer. Remember, for normally aspirated engines this probably means applying full throttle. Turbocharged engines usually require slow phase-in of power.

Some pilots insist on locking-up the brakes and applying full power before starting their takeoff roll. There is really no advantage to this at all. To become fully efficient, a propeller must be moving forward through the air, not just rotating in place. You can damage the propeller, the undercarriage, and the horizontal stabilizer during static runups on gravel strips. The prop blast will cause small pea gravel and debris to be sucked up and thrown back onto the aircraft.

Rolling Starts. You can gain a little extra speed by making a rolling turn onto the runway from the taxiway, but there are some potentially dangerous side effects to this maneuver. Fuel starvation, for example, may occur. During a rolling, turning start, fuel is thrown to the outside of the tanks by centrifugal force. This problem is exacerbated on high-speed turns by aircraft with minimal wing dihedral.

Another problem with a rolling start is that the outside main gear may collapse (this can happen on fixed gear as well as retractables). The same centrifugal force that throws fuel to the outside of the tank also imposes extra load on the outside gear. The extra load, if excessive, will cause the landing gear to buckle and collapse.

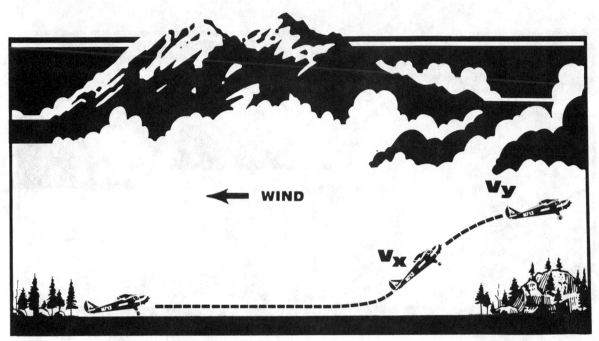

Short-field takeoff on a firm surface: Accelerate in clean configuration. Rotate at best-angle-of-climb (V_x). After you clear the obstacle, accelerate to best-rate-of-climb (V_y). V_x and V_y are critical airspeeds which every mountain flier must memorize. (Adapted from David Whitelaw drawings)

At a minimum, high-speed rolling starts may increase friction and drag on the tires which can result in a longer takeoff roll. Tailwheel aircraft are prone to ground-looping when the turn is too sharp and too fast. The best rule of thumb to minimize these hazards is to keep the ball of the turn coordinator within one ball-width of center.

The Takeoff Run. For aircraft with tricycle landing gear, keep control surfaces in a neutral position (except to compensate for crosswinds) while applying power. Lift the nose off the ground after reaching best-angle-of-climb speed (V_x). Maintain V_x until all obstacles have been cleared, then increase speed to best-rate-of-climb speed (V_y) or cruise climb, as appropriate.

For taildraggers, keep control surfaces in a neutral position until the rudder becomes effective. During the initial application of power, take care to avoid nosing over. When you reach sufficient airspeed and can maintain directional control with the rudder, lift the tail of the aircraft a few inches off the ground. If you raise the tail any higher than this, you will place excessive weight on the main gear causing increased drag and an increase in takeoff run.

On sloping runways it is best to take off into the wind and downhill. You can take off downhill with a light tailwind, but if the tailwind exceeds 10 knots, you should give careful consideration to waiting until the wind diminishes.

119

Short-field landing on a firm surface: Maintain a "steep" constant-rate-of-descent approach with power and full flaps. Maintain a constant approach speed at 1.3 Vso + ½ headwind component. Touch down no faster than 5 knots above stall speed. Reduce power after landing, retract flaps, and brake as required.

Short-Field Landings

The main objectives of a short-field landing are to maintain a constant approach speed on a constant approach path over all obstacles and to land in the shortest possible distance.

The Approach. The proper approach speed for a short-field landing is 1.3 V_{so} plus one-half the headwind component. Contrary to what some pilots profess, the approach path to a short field should be made at a steep angle and at a constant rate of descent using controlled power. The FAA recommends short-field approaches based on 1.2 V_{so} rather than 1.3 V_{so}, but we prefer the extra airspeed for an added margin of safety. If this extra speed means you're likely to overrun after landing, you shouldn't even consider landing on such a short field to begin with.

The *step-down approach* is the *wrong* approach to use for short-field landings. The step-down approach is flown lower and slower over obstacles, with a much flatter approach angle, while maintaining lots of power. After passing the obstacles, the pilot cuts the engine back to idle, drops the nose down, and lands. Not only is this a lot of extra work, it is dangerous. It requires that a pilot *depend* on horsepower to reach the airstrip. If the engine fails on approach there would be little chance of making the airstrip at an approach speed of less than 1.3 V_{so}. The flatter approach angle does not reduce overall landing distance and takes away a pilot's most valuable asset: altitude.

If your aircraft has flaps, always use them for approach and landing, but never apply full flaps until you are sure that you can make the runway. At maximum gross weight or at high-altitude airports, performance is reduced when flaps are not used.

Touchdown. The touchdown is accomplished at minimum controllable airspeed, at an attitude that will result in a power-off stall. Aircraft with nosewheels should be held in this attitude until the elevator is no longer effective. Aircraft with tailwheels should be held in the three-point attitude throughout braking.

Excessive airspeed will cause the airplane to "float" when it enters ground effect. If the runway length is critically short or no go-around is possible, a pilot could be in trouble.

SOFT-FIELD OPERATIONS

Loose gravel, dirt, grass, sand, mud, tundra, and snow fields are all considered soft fields. They dominate the landscape in remote areas of Alaska, Canada, and Mexico. In the lower 48 most soft fields are restricted to private use, but quite a few soft small-town public strips remain.

Soft-Field Takeoffs

The key to executing a soft-field takeoff lies in transferring the weight of the aircraft from the wheels to the wings. This transfer needs to be done quickly and at the slowest possible airspeed. Loading the aircraft so that the center of gravity (CG) is located near the aft section of the CG-envelope will help a great deal during soft-field takeoffs. An aftward balance aids in keeping the nosewheel off the ground with less elevator pressure. Be careful, however, not to load the aircraft beyond the CG-envelope limits.

Taxiing. In very soft field conditions, pilots may have some difficulty breaking ground. If you are unable to get the plane out of its tiedown spot, rocking the elevator and rudder back and forth while increasing power may do the trick.

Taxiing on soft terrain should be done with the elevator in the full back position. Avoid using the brakes. Many times, stopping the aircraft once it is in motion will cause it to sink down into the soft surface. Runup should also be done while the aircraft is in motion, rather than standing still, to avoid picking up debris with the propeller.

The Takeoff Run. Use whatever flap setting the manufacturer recommends. Each aircraft is different. The manufacturer's recommendations are calculated to provide the maximum amount of lift from the wing, but as a rule of thumb, flaps which are extended to the same angle as full down-aileron provide the best lift potential.

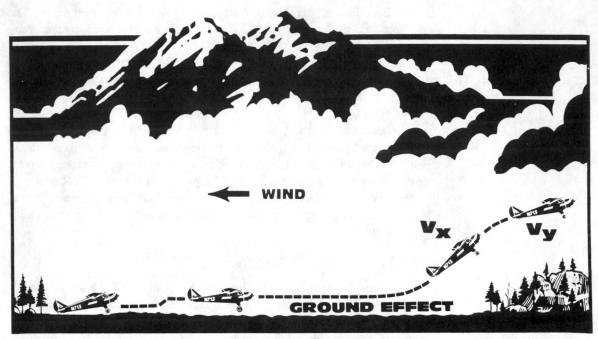

Takeoff from a short, soft field: Transfer weight from wheels to wings as soon as possible. Lift off at slowest possible airspeed. Lower the nose to stay just off the surface in ground effect. Accelerate in ground effect to best-angle-of-climb (V$_x$). Climb out at V$_x$ until you clear obstacles, then accelerate to best-rate-of-climb (V$_y$). (Adapted from David Whitelaw drawings)

If the surface is so soft that the takeoff run is difficult, rocking the aircraft with the ailerons may help the gear come off the ground more easily. Taking off from extremely *sticky* surfaces, such as mud or tall wet grass, you may need to use the aileron to lift one wheel off the ground and roll down the runway on the other. Very soft fields may decrease acceleration so much that you must abort the takeoff run.

The normal soft-field takeoff is accomplished with the elevator in the full back position. Apply full power only when the aircraft is rolling and pointed down the runway. Tailwheel aircraft should have the tail down low to the ground, no more than a few inches off the surface. Nosewheel aircraft should maintain a slight nose-high attitude, but like the tailwheel, only a few inches off the ground. Both need to rotate nose-up as soon as possible.

Once airborne, lower the nose slightly to remain in ground effect until best-angle-of-climb speed (V$_x$) is reached. Use V$_x$ to clear all obstacles, then increase speed to best-rate-of-climb (V$_y$). Flaps should be retracted only as recommended by the manufacturer.

One of the most common mistakes pilots make in soft-field takeoffs is to fly out of ground effect too soon and at too slow of an airspeed. Typically, the aircraft will settle back down onto the runway, and the pilot will find himself out of room for the takeoff attempt.

Soft-Field Landings

Especially at unattended airports, inspecting the runway before you land is highly recommended. Fly over the runway at pattern altitude (800–1,000 feet). Look for obstructions, depressions, animals, large rocks, and anything else that may interfere with a safe smooth landing. If you can't see well enough at pattern altitude then make a low approach at best-angle-of-climb speed (V_x) at an altitude of 200–600 feet. The view is much better and you're apt to see smaller obstructions. If you are still unsure about your proposed landing spot, it is simply a matter of good judgment to abort the landing and go on to your alternate.

An aerial inspection of a proposed landing site in the mountains carries with it a special warning. Be cautious of low-level wind shear and downdrafts. On frozen, snow-covered lakes, wind is usually indicated by snow drifts and clearings on the ice. Plan your landing close to the shore line if possible and practical.

Soft-field landing: Approach at 1.3 V_{so} + ½ headwind component with power. Transfer the weight from the wings to wheels gently at the slowest possible airspeed with power. Continue full back elevator, maintain directional control with rudder, and reduce power slowly. Retract flaps gently, and avoid using excessive brakes. Taxi with full back elevator, and exercise caution. (Adapted from David Whitelaw drawings)

On river bars, sandbars, and dry lakebeds, or other areas where it is difficult to determine the wind direction on the surface, pay close attention to the "natural" crab angle of the aircraft while overflying the landing site.

Most importantly, take as much time as you need to determine that a safe landing can be made. Fuel permitting, it is not at all unusual to spend 15–20 minutes looking over a landing site before deciding to land.

The Approach. Set up the aircraft before final approach. Like the short-field approach, the soft-field approach should be made with a steep glide path and at a constant rate of descent. Airspeed should be closely monitored using $1.3 \, V_{so}$ plus one-half the headwind component.

Touchdown. Transferring the weight of the aircraft from the wings to the wheels as gently as possible is the most important aspect of a soft-field landing. As the aircraft enters ground effect, reduce airspeed to approximately five knots above stall speed. Use *power* to control descent. When the airplane touches down, hold the nose high with full elevator back pressure. Taildraggers should use the conventional three-point touchdown and continue to hold back elevator pressure during roll-out.

Unusually soft surfaces such as mud, wet grass, snow, or sand will create plenty of drag to slow the aircraft. Avoid excessive use of brakes. Keep flaps down until the weight of the aircraft has been completely transferred from wings to wheels, and then retract them slowly. Raising flaps too quickly will tend to make the aircraft bog down in the surface, which may cause the plane to flip over or ground-loop.

If terrain and obstructions near the landing site don't dictate otherwise, you can test the surface of a soft field by "dragging" your wheels (or skis, if on snow) through a "balked" landing. If you do this, be sure to keep the power up high and only *lightly* skim along the surface. For less experienced pilots, this may be dangerous. Dragging the landing gear is an awful lot like doing a touch-and-go with power on. This kind of maneuver should be practiced on a hard-surfaced runway first. The technique takes a little getting used to before skills are developed satisfactorily enough to try it on a muddy or deep-snow-covered surface. If you do this maneuver improperly you could flip the aircraft over on its back, ground-loop it, or just plain "crunch" something. A "crunch" can cost a bunch.

ROUGH-FIELD OPERATIONS

Pilots from metropolitan areas are simply not used to seeing rough fields. An excursion into the mountains may be your first encounter with a field that is cluttered with sticks, rocks, brush, and other forms of debris. The more remote

Cessna landing on a gravel bar in the Brooks Range, Alaska.
(Galen Rowell / Mountain Light)

the airfields are, the less maintained they are. Extremely remote landing fields may be nothing more than a gravel bar in the bottom of some isolated river valley.

Rough-field takeoffs and landings are done exactly the same way as soft-field takeoffs and landings, with one exception. Forward visibility is an absolute must in rough-field operations. Pilots should not hold the nose up as high for a rough-field takeoff as they do for a soft-field takeoff. Nor should they land with such a nose-high attitude. Chuckholes, rocks, tree limbs, and other obstacles must be seen and avoided. A nose-high attitude on either landing or takeoff blocks forward vision.

On landing, use the forward slip to gain better forward visibility. Be prepared for sudden jolts that may occur when the landing gear bounces over rocks or falls into chuckholes. Make sure that your seat is locked so that it will not slide backwards during rough jolts. Keep seat belts fastened and tight.

Before takeoff, walk the length of the field. Clear any obstacles which may hinder takeoff. Flag chuckholes with a stick so they can be seen and avoided.

INCLINED RUNWAYS

The most common type of mountain airstrip runs up and down the mountain slope rather than along the side of the slope. Where possible, plan upslope landings and downslope takeoffs. A 1–5° slope is small to moderate; a 5–10° slope is pretty severe; and slopes greater than 10° are very tough to negotiate.

Takeoffs

Always take off downhill in light wind conditions. Acceleration is greater, less runway is required, and obstacles are cleared more easily. During takeoff, a 1° downslope is roughly equivalent to having 10% more runway; a 2½° upslope is equivalent to having a 7-knot tailwind during takeoff.

Use the short/soft-field techniques for taking off on downslope runways. Once airborne and while still in ground effect, pilots may notice that even though airborne they are still descending. The important thing to remember, however, is that the aircraft is actually climbing away from the runway surface.

Some pilots get confused about descending during takeoff on a downsloped runway. They react by pulling the nose up too much, stalling the aircraft, and crashing. On downslope takeoff it's best to disregard your vertical speed indicator.

Landings

Always land uphill unless the tailwind exceeds 10 knots. Landing downhill can be dangerous, especially where the actual runway slope parallels the approach path. Ground effect on a downhill landing may be hard to transition into and will make landing nearly impossible.

Overfly upsloped runways before landing on them. Determine the length of the runway by keeping track of your time as you fly over it. Try to determine the degree of the gradient, if possible. Use trees, bushes, people, or animals to help you decide. Avoid staring at any one object during the approach phase. Pick several items to look at and compare.

Use imaginary lines from point to point: from where the initial approach to landing begins to the threshold; from the threshold to touchdown point; and from touchdown to the runway's end.

Approach. You must maintain higher-than-normal airspeed and pitch attitude when landing upslope. This applies to both the approach and touchdown phases of the landing. As the aircraft touches down, an attentive pilot might even notice the vertical speed indicator showing a positive rate of climb. This positive rate of climb is a necessary ingredient for landing upslope, and the additional airspeed will be required to carry the aircraft into the flare.

On level terrain, an indicated approach speed of 1.3 V_{so} plus one-half of the headwind component works well, but for an upslope runway it falls short. We recommend using 1.4 V_{so} plus one-half of the headwind component for inclined surfaces. The difference between 1.3 and 1.4 is about 5 additional knots on most light aircraft. The faster approach speed bleeds off quickly during flare-out as a result of the positive rate of climb and the upsloping terrain.

The excess airspeed is a must. Quite often mountain airstrips do not afford the luxury of "going-around". Use flaps for landing, and apply full flaps only after you know you can make the runway threshold.

Two-Step Flare. With any landing there is a transition from the approach path to the flare-out for touchdown. When landing upslope, this flare transition is accentuated—it feels very much like a secondary flare. The second part of an upslope flare is necessary to hold the nose high enough above the overall natural horizon so that the longitudinal axis of the airplane parallels the inclined surface.

Without extra airspeed, the rate of descent will be too high during the second step of the flare and will require substantial power to correct. Besides, most of the time, adding power during the last few seconds of a landing is ineffective because of the time it takes for the engine and propeller to react.

Touchdown and Roll-Out. Many airplanes lack the performance capabilities to operate on upslope/downslope terrain, especially at higher elevations. On steep slopes, more than 10°, and at high density altitude, the sustained climbing rate for your airplane may well be exceeded. Remember, in order to make an upslope landing the aircraft must be flown parallel to the surface during the last few feet of the flare and roll-out. This may turn out to be a strong climb, one that taxes your aircraft beyond its capabilities.

The extra airspeed you take into the approach and flare will have little if any effect on landing distance. The positive-rate-of-climb attitude and the upsloping terrain will reduce ground speed quickly.

The downhill takeoff: Use the short/soft-field takeoff technique. Because of the increased angle of attack during the takeoff roll, the aircraft will become airborne very quickly. Once airborne, lower the nose to stay in ground effect until best-angle-of-climb speed (V_x) is attained. Climb out at V_x until obstacles are cleared, then accelerate to best rate of climb (V_y). Remember that the aircraft will have a positive rate of climb in relationship to the runway slope but a negative rate of climb in relationship to the horizon. Always fly the airplane with relationship to the runway. Watch the surface, not the horizon! During the departure, fly by reference to the descending terrain until clear of all obstacles and terrain features. Always take off downhill unless the wind is more than 10 knots downhill. (Adapted from David Whitelaw drawings)

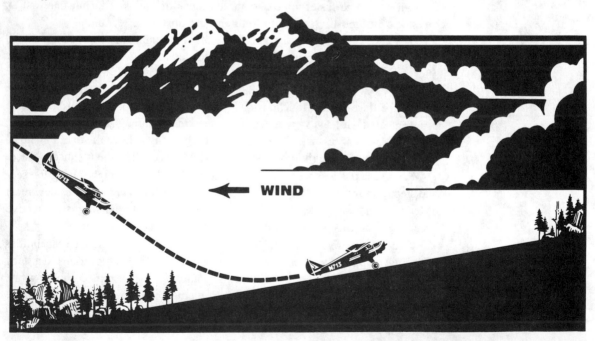

Rising Terrain, Upslope Landing: Maintain steep constant-rate-of-descent approach with power. Use full flaps only when you can make the runway environment. Maintain $1.4\,V_{so}$ + ½ headwind component. Within ground effect, increase power to decrease rate of descent. Adjust pitch attitude with elevator to parallel the longitudinal axis with the runway. Touch down within 5 knots of stall speed. Retract flaps, and brake as required. (David Whitelaw)

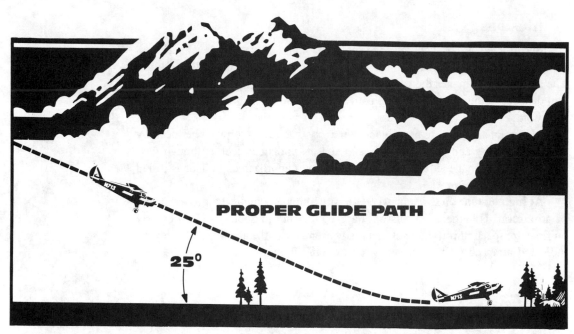

Approaching an upslope runway, don't be deceived into thinking your glide path is too steep. You'll need the extra height to clear obstacles in front of the threshold. (David Whitelaw)

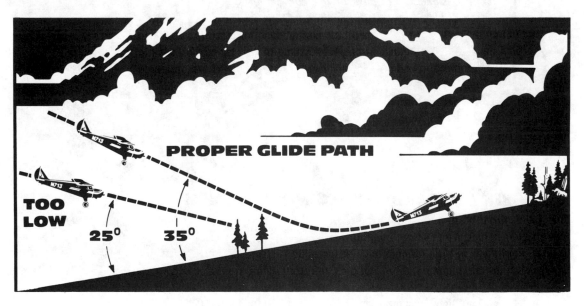

Illusions

Because you are probably used to shallower approach paths to level surfaces, an approach to an upsloped runway may give you the illusion of being too high. The steeper angle makes the runway look closer and shorter. All too often pilots will descend below the proper glide path and touch down too short of the runway threshold.

Mountain airstrips are usually more narrow than urban runways. Especially in poor visibility and during sunrise and sunset, pilots can feel as though they are too high on final and descend below the proper approach path. This is called *runway width illusion*.

At high-altitude airstrips or when density altitude is high, the ground speed of an aircraft is higher. Objects on the ground seem to zoom by. Particularly on final approach, throttling back can inadvertently stall the airplane. Watch your indicated airspeed—it's what keeps the airplane flying.

*Both strips are the same length.
Runway width illusion causes you
to think you're too high on approach,
especially at dusk or in poor visibility.
Don't let this phenomenon persuade
you to descend too low.*

TAKEOFFS AND LANDINGS—POINTS TO REMEMBER

⚠ The higher the gross weight, the higher the takeoff speed, the longer the ground run, and the higher the stall speed.

⚠ An improperly loaded airplane, with its center of gravity out of limits, may not be able to be controlled.

⚠ Landing "light" can mean landing long. If you fly out to your fish camp, unload passengers, baggage, and fishing poles, and return alone with low fuel, you can expect to "float."

⚠ High density altitude will decrease performance of normally aspirated engines, decrease propeller efficiency, and increase true airspeed, takeoff runs, and landing rolls.

⚠ At high density-altitude airports your engine will not develop its maximum rated power for that altitude unless it is leaned to the proper fuel-to-air ratio. This can be particularly important if a "go-around" is necessary.

⚠ Gusty wind requires you to keep the airplane on the ground longer during takeoff.

⚠ Crosswinds will require more airspeed during the approach and touchdown. The higher the airspeed, the smaller the wind correction angle necessary, but the more runway you will need to land the aircraft. In the mountains, terrain often limits runway length to bare minimums.

⚠ Know the maximum crosswind component for each aircraft you fly.

⚠ Flap settings are not standard for all aircraft. Flaps should be used only as recommended by the manufacturer. Exercise caution when performing slips with flaps.

⚠ A key approach speed for short-field, soft-field, and crosswind landings is: $1.3 \ V_{so} + \frac{1}{2}$ HWC (Headwind Component).

⚠ Do not count on a headwind to decrease landing roll-out unless it exceeds 10% of your touchdown speed.

⚠ Normal approach speed should be increased whenever the headwind component exceeds 5 knots. A good rule of thumb is to divide the headwind component of the reported maximum gust by two and add the result to your normal landing speed:

$$1.3 \ V_{so} + \frac{1}{2} \text{ Headwind Component of Maximum Gust}$$

Normal Approach Speed + Gust Factor

⚠ Takeoff distances included in aircraft operating manuals are predicated on ideal conditions: a paved, dry, level runway.

⚠ Braking becomes more effective as airspeed and lift decrease. If your plane has flaps, retracting them just after touchdown will decrease lift and put more weight on the landing gear, but don't mistakenly raise the gear instead.

⚠ Ground effect makes it possible for an airplane to lift off too soon with an excessive high pitch angle, an excessive load, or both, increasing your chances of stalling. You may not be able to accelerate to climb speed without first lowering the nose.

⚠ Airspeed is the most important factor in executing a precision landing. Airspeed control begins in the traffic pattern. You should know and use the appropriate V-speeds for each segment of your approach. Except in conditions where strong winds are present, an aircraft on short final with wings level should use a speed of $1.3\ V_{so}$ when landing.

⚠ Roll-out distance can be calculated by dividing the actual touchdown speed by the normal no-wind touchdown speed, squaring the result, then multiplying by the normal no-wind landing distance.

⚠ Emergency planning is a must in preparing for takeoff. Most power losses occur during the initial application of power. Don't rush your runup before takeoff. Use a takeoff checklist. Remember, if you lose power, maintain airspeed and control the aircraft at all times.

⚠ On normal landings where terrain permits, the best accident prevention maneuver is the go-around.

⚠ The most common errors pilots make in executing a smooth take-off or landing are: failing to direct their vision properly; failing to control their heading on the ground properly; poor planning; and poor airspeed control. In the mountains, airspeed control is crucial.

6
Floats and Skis

THIS CHAPTER IS NOT INTENDED TO BE A DEFINITIVE TREATISE ON FLOATPLANE or skiplane operations. Rather, our purpose is to highlight these operations from the perspective of the mountain flier. More complete instructions for floatplane flying can be found in *Flying a Floatplane* by Marin Faure (TAB BOOKS, 1985). Consult back issues of *Alaska Flying* for detailed articles on skiplane operations.

FLOATPLANE OPERATIONS

Unless you are forced down over water (ha!), you must hold a current seaplane rating to make takeoffs and landings from the water, and your plane must be specially equipped for water use.

Floatplane flying requires extra planning. During preflight planning make doubly sure you can get into and (more importantly) back out of "water ports." It is possible, for example, to land on a high-altitude lake but not be able to take off again. Check the performance capabilities of your aircraft. Water surfaces have a great deal more drag than asphalt runways, so your plane will need more distance to take off. If you combine high density altitude with the additional drag, the result is an even more extended takeoff than for a normal land takeoff. Be wise. As with short fields, clock the length of the water port from end to end to ensure that the distance for takeoff is there *before* you land.

Taxiing

When floats are mounted on an aircraft, nearly all of its normal shock-absorbing ability disappears. It is possible to damage an aircraft by taxiing too fast across rough water. Taxiing on the "step" should be kept to a minimum when the water is rough.

Step Turns. There are several advantages of step turns. They will greatly reduce the required takeoff run in small lakes, and they are handy for turning an aircraft from an upwind to a downwind position. While water rudders work well for turning in wind speeds of less than 20 knots, in stronger winds it may be impossible to turn the aircraft from upwind to downwind with the use of water rudders. The step turn will be helpful. The wind will not flip the aircraft because it is opposed by the centrifugal force created during the turn.

Care must be taken during the step turn so that the aircraft does not become airborne. This is best done by keeping the speed of the aircraft down so that the back of the float step is used to support the weight of the aircraft. This will keep the center of hydrodynamic pressure aft. The approximate speed difference between too fast and too slow in a step turn is approximately 5 knots. Too fast and the plane wants to fly; too slow and it will porpoise on you. If porpoising starts, add a bit of power and try increasing the back pressure on the stick. Step turns should not be done close to the shoreline. It may be necessary to decrease the rate of turn, and without sufficient room the floatplane might end up on the shore or in the trees. Allow plenty of room.

Step turns from the downwind to the upwind position are not recommended. The centrifugal force created will aid the wind in flipping the aircraft over or tipping a wingtip in the water. Do not attempt this type of turn.

Takeoffs

Determining if the length of a takeoff area is adequate, once on the water, can be accomplished by the following method:

Taxi at idle engine power over the length of the selected takeoff run. Let's say that it takes 5 minutes or 300 seconds. Place a small empty biodegradable bottle in the water at approximately 62% down the length of the runway as a "go/no-go" marker. To do this, back-taxi to the starting point again, and slow-taxi up the intended takeoff run for approximately 180 seconds (62% × 300 seconds ≈ 180 seconds). Toss out the bottle, and taxi back for takeoff. If you are unable to attain approximately 75% of your takeoff speed at that point, retard the engine and abort the takeoff. This procedure works well for taking off from small lakes with obstacles.

You should also get used to timing your takeoff runs. Once you know your average time to break water, you can use that as a guideline for establishing a "go/no-go" decision point based on elapsed time.

Water rudders should be used only for water taxiing to and from the dock, not for high-speed operations like takeoffs and landings. Fine pitch-attitude and aileron control is the best way to get up "on the step" and off the water. You will find that the right float will break away from the water easier than the left one, due to engine torque.

Small Lake Takeoff. When winds are greater than 5 knots and small waves are present, it is often possible to make a direct takeoff from the downwind end of a small lake to the upwind end. The headwind enhances your airspeed and the waves help you get up on the step and break away early.

When winds and water are calm, however, it may take time to get on the step, and breaking away may be difficult. In these conditions, use the technique depicted in the accompanying diagram. Be sure to monitor your power to avoid becoming airborne too early, and don't add full power until you are lined up for takeoff. The wake generated during the step turn helps you break away from the surface easier, reducing the water run. Practice this technique on large lakes before attempting the real thing.

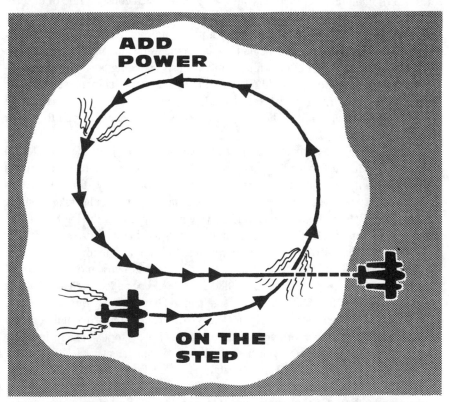

Small lake takeoff. Get "on the step" early, add power as you begin your "base leg," and break away as you cross your wake.

After takeoff, obtain best angle-of-climb (V_x) and begin a standard-rate turn to increase the clearance between you and the obstacles on shore. The turn will not significantly reduce the amount of lift because the angle of bank is shallow, and the distance covered in the turn translates to as much as 100 feet of extra altitude over the trees.

Landings

When landing on a lake, it makes good sense to overfly it first, whether you've landed there previously or not. Debris, like floating logs, barrels, and such, can be lying just beneath the surface of the water and not be visible to the pilot until it is too late.

Check wind direction. The upwind side of a lake will normally have calm water while the approach end will be slightly more choppy. Land into the wind, and bring any drift under control before touchdown.

Water landings are normally power-on landings (although you should occasionally practie emergency power-off float landings). Use full flaps unless the aircraft manufacturer instructs otherwise. Establish a constant rate of descent of 200 feet per minute well in advance of touchdown. Keep power on all the way to touchdown. Control your descent with power and your airspeed with pitch. As the floats make contact and begin to settle on the water, increase back pressure to full up-elevator, and hold it while simultaneously reducing power. Like a soft-field landing, touchdown should be gentle. Drag on the floats increases with the square of the speed, so excess speed will produce a staggering increase in the load on the floats and hull. When the plane settles in the water, extend the water rudders and retract flaps.

If you were to ask most experienced floatplane pilots how they would tell if a lake is suitable for landing, they would tell you that if the lake has cabins or a dock on it, more than likely it is landable. Unfortunately, this still leaves a bit of doubt as to what *type* of aircraft can get into that particular lake. Timing the lake when flying both upwind and downwind, then taking the average is the best way to determine if the lake is long enough for the departure.

Let's say that at 60 MPH (IAS) it takes your heavily loaded Cessna 185 approximately 20 seconds to fly the distance. This translates into 1760 feet of usable water run. Another 10 seconds (880 feet) should be added to clear a 50-foot obstacle, making 2640 feet. This is a rough method but it's fairly quick and easy. Check the float operations supplement in your aircraft flight manual for data regarding maximum performance takeoffs to clear a 50-foot obstacle in order to determine if the distance you have calculated is sufficient.

Crosswind Landings. Approximately ⅓ of all float-related accidents are caused by allowing the center of hydrodynamic pressure to move too far forward during the landing. Of those accidents, the majority of pilots were better versed with taildraggers. In a typical scenario, the pilot touches down with the upwind

A high mountain lake in Pakistan. Pilots land on some lakes only to discover they can not takeoff from them because of the altitude. Helicopters must be called in to retrieve the aircraft. (Galen Rowell / Mountain Light)

float first to counteract the drift. After touchdown he attempts to plant the downwind float on the water by applying forward pressure on the yoke (which is standard procedure for conventional-gear aircraft) and then adds aileron to lower the downwind float onto the water. By applying the forward pressure he moves the center of hydrodynamic pressure dangerously forward, causing the toe or side of the upwind float to catch the water, bending a wingtip or cartwheeling the aircraft across the lake.

Don't force the high float down onto the water. Let it settle by continually applying upwind aileron as the aircraft decelerates. Increase back pressure and allow the high float to settle in the water on its own. Use the rudder control to maintain heading on the surface by picking an object on the far end of the lake.

Glassy Water Landings. The glassy water landing is done in the same manner as the featureless terrain landing to be discussed later in this chapter. Set up your rate of descent at 200 FPM. Some pilots prefer to use 1.15 V_{so} for the entire approach and touchdown. We prefer to maintain 1.3 V_{so} until passing treetop level, and then reduce power to 1.15 V_{so} while maintaining the 200-FPM descent. The extra airspeed increases the odds of reaching the water, while still under control, in the event of engine failure.

As the aircraft enters ground effect, the rate of descent will decrease by as much as 100 FPM (check your vertical speed indicator if you don't believe us). This change, when noted, will tell the pilot that the aircraft is near the surface. The nose should be held near level with the far end of the lake, and touchdown will follow. Once on the water, reduce power, gradually increasing the back pressure to full up-elevator, and retract flaps. After the aircraft is settled in the water, extend the water rudders for docking.

The power-on approach is particularly important for a gentle landing on glassy water—you may not even know when you touch the surface.

Prior to attempting these landings you will need to experiment with your aircraft to determine which power settings will achieve the desired airspeeds and rate of descent.

Rough-Water Landings. You can damage your floats if you make a rough-water landing too fast or at too high of a pitch attitude, like in a full-stall landing. Keep power on until the floats touch the water. Consult the float manufacturer's recommended procedures. We prefer to touch down at a slow airspeed with full flaps and the back of the main hull touching the water surface first, increase back pressure, cut the power, retract flaps, and let the aircraft settle off the step as quickly as possible.

River Operations

Rivers are typically rougher than lakes, and most rivers have many bends, making it difficult to find a straight section that is long enough to land on and depart from. Also, many small aircraft are simply not capable of enough power to handle swift currents.

There has always been controversy as to which way to take off and land on a river, upstream or downstream? The best rule of thumb is to take off and land so as to minimize the speed of your floats relative to the water. This ensures less abuse to the floats and better takeoff performance, and means that you should usually take off, and often land, downriver. Of course, terrain features or strong winds might leave you no choice in the matter.

Although river approaches are similar to lake approaches, you will usually use the rough-water landing technique during the flare and touchdown.

When docking, turn the aircraft upstream, and using rudder and power, taxi sideways toward the shoreline.

River operations can be especially tricky during docking or beaching on shore. Because you may need help from a passenger when you jump from the float to the shore, give that person exact instructions on departing the airplane so he will

One of the hundreds of floatplanes docked at Lake Hood Floatplane Base in Anchorage. Lake Hood is the largest floatplane base in the world. (Steve Woerner)

not get tangled up in the propeller. Never allow a passenger to move in front of the wing strut, and when departing the aircraft have your passengers walk behind the wing as they step ashore. Everything happens very fast with river operations, and it is possible to forget to turn off the mags. If your passenger decides to grab the propeller for a handhold, it could cost him his life.

You must receive special instruction to be safe and proficient in river operations.

SKIPLANE OPERATIONS

Skis open up a realm of mountain flying that is off-limits to pilots of wheeled aircraft—snowfields, frozen lakes, even glaciers. Many fixed tricycle-gear aircraft and virtually all taildraggers have been approved for skis. Because skis affect weight, balance, airspeed, and other performance characteristics, they require

Cessna landing on Kahiltna Glacier, Mt. McKinley, Alaska. (Galen Rowell / Mountain Light)

a Supplemental Type Certificate (STC) before they can be mounted on an aircraft. They must be installed by a certified A & P mechanic and cannot be installed by the pilot.

The type of ski you use will depend on the type of aircraft you have. The Landis 2000 straight ski, for example, is approved for small, light aircraft like Super Cubs, while the Federal 3600 wheel ski is approved for heavier workhorses like the Cessna 180 and 185. Check with your local A & P mechanic about the type and size approved for your aircraft.

Preflight and Taxi

Skis deserve special attention during preflight inspections. You must check nuts, bolts, fittings, cables, and bungee cords before taxi and takeoff. During the winter, bungee cords get wet and freeze, causing them to fatigue, crack, or simply fail. Metal springs are highly recommended in place of bungee cords.

Taxi carefully on skis. Except with wheel penetration skis, you'll have no brakes. Direct all of your attention to the outside of the aircraft, and taxi at walking speed. Looking for maps or checking mixture, magnetos, or carb heat can put you into a snow drift or another aircraft. Like driving a car on ice, anticipate the required stopping distance when taxiing behind another skiplane.

Takeoff

Takeoff is generally easier with skis than with wheels because there is less friction; tires grab more side load. Use the soft-field technique for takeoffs with skis. With tricycle-gear airplanes, allow the nose gear to fly off the surface. If you fly a taildragger, be sure to keep the tail on the surface until you can maintain directional control with the rudder.

Landing

Again, use soft-field techniques. Allow the heel of the ski to touch down first. Touch down with full flaps and as slow as possible, unless the manufacturer recommends otherwise. After touchdown, keep the elevator in the full back position and reduce flaps. For tricycle-gear aircraft, continue holding a little back pressure on the elevator while you taxi to parking. For conventional gear, the tailwheel will help slow the aircraft during the landing roll.

Frozen Lakes

Loose water, or *overflow*, caused by water seeping up through the ice (or sometimes by snow melt), is a real problem for skiplane operators. Locating and staying out of overflow areas is easy if you know how to recognize where the overflow is likely to be. Remember, water always drains to the lowest point on any type of surface. Normally the center of a lake will have the worst overflow.

Doug Geeting's Cessna skiplane in Denali National Park, Alaska. (Barbara Cushman Rowell / Mountain Light)

Inlets and outlets of lakes should also be avoided. And if small amounts of fog are present on the lake, a pilot can expect overflow in that area of the lake.

Overflow can be detected by its greenish color. If overflow is a problem, take off or land close to shore, keeping in mind, of course, that if you are too close to shore you may encounter mechanical turbulence.

If you get into an overflow situation, ice will likely form on your skis. It must be removed before you attempt to take off. Jack the skis up using a log or tree limb as a lever, or skid the skis up on small logs or two-by-fours, and remove the ice from the ski bottoms. If there is quite a lot of overflow, it may be necessary to "pack-down" a runway out ahead of the skis. Packing the snow in an area of overflow allows the water to freeze more quickly when it comes in contact with the air.

Featureless Terrain Operations

When drifts or mounds of snow accumulate around runways, they may interfere with ground references for landing. These drifts can alter the apparent shape and size of a runway. Familiar trees, poles, or buildings, for example, may be dwarfed by snow piled around them. Snow can completely cover or re-contour the earth's surface. In remote flat country or low brush mountainous areas, reference features may be entirely gone. The result is that your depth perception is severely hampered as you set up for final approach. The absence of sunlight, or diffuse light through a cloud cover, can aggravate this problem even further. Without shadows, a pilot has no reference point at all.

Featureless terrain is not limited to snow and ice cover. Deserts, glassy lakes—even expansive meadows—can also confuse your eyes. To examine the effects of featureless terrain, do this simple experiment. Take a piece of white bond typing paper. Wad it up into a ball, then unfold it and place it flat on a table. Now take a lamp and hold it over your head at a 45° angle to the unfolded paper. Notice that shadows appear at the ridges and creases. Now hold the lamp directly over the top of the paper, and notice how all the shadows disappear. When you turn the lamp off (to simulate night or an overcast), it's twice as hard to distinguish smooth areas from rough areas.

Try to visualize yourself on final approach to that piece of paper. What would you look for to gauge your altitude? If features are missing, covered, or obscured, the best reference you have is the altimeter. If possible, set the altimeter to the current FSS setting. In remote areas, setting your altimeter to known mountain peaks will be helpful. USGS charts reference elevations more thoroughly than Sectionals and WACs.

It is vital that you overfly and visually inspect featureless terrain before attempting to land. If there is any doubt concerning the condition of the surface, fly on to an alternate.

Approach. Approaches to featureless terrain are made easy by timing the approach. Sometimes called a *90°-270° course reversal*, this visual procedure somewhat resembles a timed nonprecision instrument approach.

Once you know your altitude, fly directly over your intended landing area *downwind* at 1,000 feet above the surface. Pick a touchdown point and start timing yourself for two minutes beyond that point. Take note of your downwind heading.

At the end of two minutes, depending on the surrounding terrain, turn 90° right or left of the downwind heading. When you reach the 90° point begin a 270° turn in the opposite direction of the 90° turn just completed. This maneuver will leave the aircraft pointed upwind for final approach and the touchdown spot you selected, but it's not foolproof. Keep in mind the effects of the wind.

Now, on final approach and at an altitude of 1,000 feet, set up an approach speed of 1.3 V_{so} plus a wind factor equal to one-half the headwind component.

143

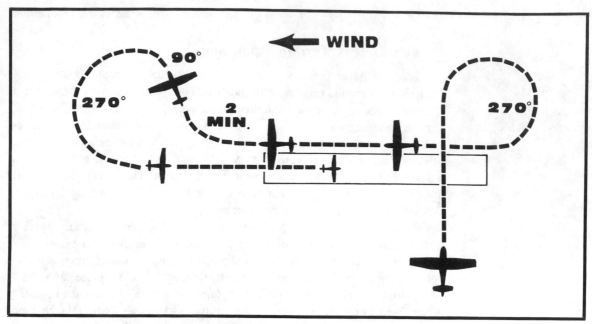

Featureless terrain 90°-270° course reversal.

(Some pilots prefer to fly the entire final approach at 1.15 V_{so} plus the wind factor, but until landing is assured—at an estimated 200 feet above the surface—we prefer 1.3 V_{so} plus the wind factor for an extra margin of safety against a stall.) Your initial rate of descent will be dictated by winds and surrounding terrain. Typically it ranges from a shallow 200 FPM to 600 FPM for a steep approach.

At 200 feet AGL (based on your best estimate), deploy full flaps, increase power, and reduce speed to maintain a 200-FPM descent at 1.15 V_{so} plus wind factor. The resulting pitch attitude should be maintained, with the help of trim, all the way to the surface. Use power to control rate of descent and pitch to maintain airspeed.

Since ground effect decreases the induced drag on an aircraft, the rate of descent will decrease slightly without adjusting either pitch or power. Touch down just above stall speed with power on. After the wheels or skis touch, hold the attitude of the aircraft constant, while reducing power and retracting flaps. Taking the flaps off will ensure the aircraft stays on the ground after landing.

Some Common Errors. The most difficult aspects of an approach to a featureless terrain landing are holding precise heading, maintaining proper airspeed, and maintaining an exact rate of descent. Concentrating too much on the instruments can make you lose sight of what few ground objects there may be.

Practicing the course reversal maneuver often will help you visualize your position relative to the landing threshold and will enhance early recognition of the changes caused by ground effect.

144

Special Considerations. In limited lighting or at night, standard-rate turns, rather than steep turns, provide a better degree of safety, especially when operating near the ground. Pilots can easily become spatially disoriented when exercising a 90°-270° course reversal if the bank is either too steep or too slow.

The bank angle in a standard-rate turn is governed by airspeed. The faster the airspeed, the greater the bank angle necessary to hold an aircraft in a standard-rate turn. Generally speaking, 15% of airspeed equals bank angle:

True Airspeed (MPH)	Angle of Bank (3° per sec.)
80	12°
100	15°
120	18°

Glacier Operations

If you fly in the Pacific Northwest, Canada, or Alaska, you are in for a special treat. Incredible from the air, ice fields and glaciers are some of the most spectacular sights a mountain flier will ever see. There are an estimated 100,000 glaciers in Alaska alone, covering nearly 29,000 square miles. Although seemingly inviting, these areas are dangerous.

Glaciers are formed when the amount of snowfall exceeds the rate of melt. Glaciers are literally rivers of ice, constantly moving and shifting. The movement of these giants is usually imperceptible, but they have been known to surge forward at a rate of more than 100 feet per day (Hubbard Glacier, Southeast Alaska, 1987).

A thick blanket of snow or crust generally covers the middle to uppermost portion of a glacier. Beneath this crust are deep cracks called *crevasses*. They generally run perpendicular to the overall ice flow and can be anywhere from a few inches to several hundred feet wide and up to 600 feet deep. The crevasses may be exposed only at the lower end (terminus) of the glacier, but they extend like ripples all the way to the upper end (head) of the glacier. The season of the year affects the strength of the crust. During the warmer months, the crust is often too weak to support the weight of a man, let alone the weight of a fully loaded aircraft. Mountain climbers tie themselves together with rope in order to avoid falling through the crust into the deep crevasses below.

Sastrugi is the snow drift formed by high winds blowing across the surface of the glacier and resembles a sand dune. *Ice fall* is an unusually steep slope in a glacier's course, down which ice flows with such high velocity that it becomes very badly crevassed. Even though a glacier's movement may be undetectable,

Overleaf: Sastrugi, Banner Peak, High Sierra, California. (Galen Rowell / Mountain Light)

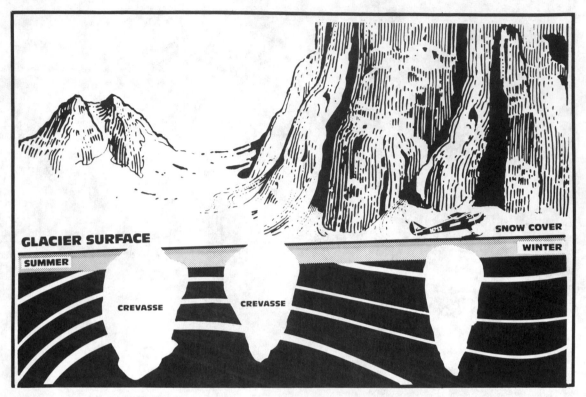

Cutaway view of a glacier and its crevasses in summer and winter. (David Whitelaw)

in an ice fall the ice is actually moving quite a bit faster than the rest of the glacier. Consequently, steep slopes and deep crevasses form. *Moraine* refers to the
edges of a glacier, as well as its terminus (called *terminal moraine*). Moraines
are distinguished by debris such as rock and huge deposits of dirt, and are often
covered by dirty snow or black ice.

Although a special pilot rating is not required for glacier operations, we feel
much training is needed. We do not recommend landing on a glacier unless you
have received specialized flight instruction. Because the snow-covered crevasses,
sastrugi, and flat light associated with glaciers are extremely dangerous, we would
classify them as unsuitable landing sites, especially for the inexperienced mountain
flier. In the interest of safety and for those who wish to view these beautiful
wonders, we offer the following *abbreviated* landing and takeoff procedures.

Glacier Landings. Glaciers are usually located at high altitudes. Mixture
should be set and left at the proper setting for the glacier elevation. If you are
on skis (while it is possible to land on a glacier without skis, skis are required
to avoid damaging the aircraft) and must make a glacier landing, treat it much

the same as you would a *very soft-field* landing. That is, overfly the surface first; look for crevasses—notable lines that run perpendicular to the overall flow of the glacier. If you can see these lines, *don't land near them*. Rather, pick a spot between the lines. Check the slope of the glacier. Pick the longest, smoothest, flattest stretch you can find between the buried crevasses. Unless there is a strong tailwind, glacier landings are made upslope. Pick a spot that will give you an upsloped runway.

Check the depth of the snow. A black plastic garbage bag with a few potatoes or a couple of cans of spam in it works best. (If you get stuck, you can eat the contents.) Toss the bag out of the airplane from about 500 feet. If the bag disappears, the snow is too soft and too deep to land. If most of the bag sinks below the surface but a portion remains above the surface, a landing is possible but it will be very soft and will require a lot of shoveling and snowshoe work to pack down a runway for takeoff later. If the bag bounces off the surface, the crust is iced over and a safe landing is possible, but don't expect the crust to be as hard as asphalt or concrete.

Then, define a touchdown zone by dropping another bag in line with, and upslope from, the first. Flat light eliminates depth perception, so the bags not only help define the touchdown zone but also provide a degree of depth perception. The bags create a kind of "poor man's VASI." If the bags appear to move farther apart on approach, you are too high; if they move closer together, you are too low. If the farther bag appears to be to the right of the closer one, you are to the right of the course; if it appears to be to the left, you are to the left of course.

You must find the flattest spot. If there is a small hill or sastrugi off to the side of the runway selected, you must adjust the approach accordingly. Landing on a side of a hill will cause damage to the wing and will probably result in the upslope ski becoming stuck and flipping the aircraft over on its back.

Keep the wings level with the runway cross-slope on approach; in other words, follow the lay of the terrain. This probably will not agree with the aircraft's artificial horizon. Final approach and touchdown are performed just like any other soft-field approach and landing. Check density altitude. You may be surprised to find that the surface temperature on top of a glacier is much warmer than you might expect. Check crosswind against crab angle. Be alert for low-level wind shear and severe updrafts/downdrafts. Apply full flaps, and check that the skis are down.

Overleaf: Doug Geeting on Ruth Glacier beneath Mt. Dickey, Alaska.
(Galen Rowell / Mountain Light)

FLOATS AND SKIS

Final approach is the same as for a soft field: full flaps, controlled descent, flare, power off, touchdown. Many glacier pilots find that a two-step flare works better than a normal flare—as the aircraft begins to settle in its original flare, bring in 75% power to produce a secondary flare and added cushion. Keep flaps in the full extended position all the way through roll-out, and maintain directional control until the aircraft has come to a complete stop. After landing, the airplane should still be faced uphill.

Glacier Takeoffs. Glacier takeoffs should be made in the same tracks made during landing, only *downhill*, towards the glacier's terminus. If the snow is soft (deep), you must pack or shovel the snow immediately in front of the aircraft. Make a looped trail so that the aircraft can be aligned with the landing tracks.

Below: Doug Geeting (climbing out of the plane) on Mt. Foraker near the Kahiltna Glacier, dropping off mountain climbers. (Steve Woerner)

The softer the snow, the more drag. If the snow is too soft, a pilot may have to "pack down" a runway in order to take off.
(Steve Woerner)

Packing is dangerous—you could fall through the crust into a crevasse. Do the same as mountain climbers: tie yourself off with a rope while packing. If you have no passengers, tie the rope to the aircraft. Falling through the crust into a crevasse would mean *death*. Don't take the chance.

Complete a standard preflight inspection and runup first. As with other soft-field takeoffs, departing from a glacier requires a quick transfer of weight from the wheels to the wings. Lower the flaps for maximum lift. Set the mixture for the altitude of the glacier. Apply steady and continuous power. Once on the roll, avoid stopping in the snow. Stopping may result in the aircraft sinking into the surface. If you must stop and cannot get the plane rolling again, try full power and rock the airplane up and down with the elevator control. It is best, however, to make every effort to keep the aircraft moving.

Maintain directional control in the landing tracks as much as possible. Getting out of the tracks will cause a sudden increase in drag and may result in insufficient airspeed for takeoff. Like other soft-field takeoffs, after the craft is airborne and while it is still in ground effect, lower the nose briefly and accelerate to best-angle-of-climb speed (V_x), then retract flaps and accelerate to best-rate-of-climb speed (V_y).

Because glacier takeoffs are downhill, you may notice that you become airborne even though the vertical speed indicator indicates a descent. Don't be fooled by this indication; you are still airborne and, in effect, climbing. Just be sure to monitor your airspeed.

7

Emergency
Procedures
and Survival

THE BEST ADVICE TO MOUNTAIN FLIERS IS TO BE PREPARED FOR EMERGENCIES. Engine failures, in-flight fires, electrical problems, icing, propeller overspeed, and landing gear problems do happen. When they happen in the mountains the emergency is compounded. Most emergencies are caused by a lack of careful preflight planning and could be avoided. Familiarize yourself with the emergency procedures outlined in the owner's handbook for the airplane you fly, and be prepared to take quick decisive action. Preflight yourself, your aircraft, route of flight, and weather.

Some incidents, however, are unavoidable, or at best unable to be anticipated during the most thorough preflight. Perhaps the most disarming emergency is an engine failure—anywhere, anytime. When it happens most of us get a ''rush.'' We get anxious and stiffen up, no matter how many hours we have under our belts. Panic, however, can turn a relatively simple emergency into a full-blown disaster in a very short time. Being in the mountains makes matters worse. Landing sites of any type can be a scarce commodity in rugged terrain, in a place where jagged cliffs plunge into a river bed below, filled with trees and rocks the size of cars. Give some thought to emergencies *before* you take off. You won't find any of those perfect little flatland cornfields in the kind of territory we've been talking about.

During preflight planning, mark your chart clearly with a highlighter pen.

Note those places you think might make suitable landing sites in the event of an engine failure or other emergency. Once airborne, don't use an autopilot. Pay attention to what you are doing and where you are going. Keep a constant vigilance for potential landing sites should the engine quit in flight. "Keep your eyeballs peeled." In the mountains, it's a good practice to "fly from one emergency landing site to the next." Landing sites may be anything from a hilltop meadow to a tiny clearing on the side of a mountain, to a sandbar in the middle of a river. In a real emergency these kinds of sites may be all you have to choose from.

LOST PROCEDURES

Becoming lost or disoriented is not in itself a life-threatening situation, but the fear and anxiety of being unsure of your position can turn this simple problem into a real emergency.

Poor visibility, caused by rain, dust, smoke, smog, snow, and the like, contribute to disorientation. These conditions can cause the most experienced pilots to lose their sense of direction. Low clouds, particularly over featureless terrain, seem to aggravate disorientation. Or, poor reception of radio signals in mountainous terrain may be the culprit. Whatever the cause, it is easy to get lost.

Pilots need only exercise a few precautions to safeguard against this problem. Studying charts and maps will help tremendously. Take note of rivers and creeks. Do they flow north or south, east or west? At what points on the map do they change or curve? Remember, water always flows downhill.

Pinpoint towns and settlements, villages, even small cabins indicated on your map. Towns and villages tend to be located along navigable rivers and waterways. Maintain a list of landmarks that *must* be found along your route, and check them off as you pass.

Experienced pilots who encounter poor visibility or simply become lost will look for rivers, streams, powerlines, mountain peaks, ridgelines, or anything that will give them a clue as to where they are. Knowing that rivers flow downhill gives them a direction to fly. Chances are, further downriver they will fly over an identifiable reference point.

Trust your compass. It's the best instrument to get you out of trouble—if you correct for variation, deviation, and wind. A triangular pattern or a timed zig-zag pattern might be fine if you have unlimited fuel and are lost over civilization, but don't waste valuable fuel doing those over remote mountain areas. In the mountains the best method is always one which takes you towards lower terrain, away from peaks and ridges, and towards a town or settlement.

Keep track of your last known checkpoint. If worse comes to worst, simply fly a heading that will take you back to that checkpoint.

And most of all, don't hesitate to call for help from an FSS, ARTCC, UNICOM, military base, or anyone else within radio reception.

CRASH SURVIVABILITY

The most important aspect of any emergency is that you and your passengers survive. Almost any type of terrain can be survivable in a crash landing—even the side of a mountain—as long as you know how to use the aircraft's structure to cushion the landing and protect the occupants.

When a crash seems inevitable, remember, it is much better to have the impact occur straight on, from the front, rather than from the side. Pliable or frangible objects like wings, landing gear, tail section—even the belly of the aircraft—can be used to absorb energy for a gradual deceleration.

Restraining Systems

The importance of wearing approved safety harnesses and belts, and keeping them properly adjusted, cannot be overstressed. Without them your chances of surviving a forced mountain landing are next to nil.

Deceleration time can be less than three seconds and still be survivable if the seat belts and shoulder harnesses are on tight. For example, the speed of an F-4 Phantom when touching down on a carrier deck is around 130 knots. The F-4 hits the arresting cable and decelerates to zero in three seconds—and in less than 400 feet.

An aircraft that touches down at 60 knots can come to rest in three seconds in a distance of 150 feet. The forward horizontal g-loading would be about 2 g's—quite survivable but without seat belts and harnesses, severe bodily harm would be likely.

If your aircraft still has the old cloth-to-metal belts, take them out and throw them away. Replace them with metal-to-metal belts. Worn-out belts should also be replaced. Slamming the cockpit door on any belt will weaken the webbing.

EMERGENCY LANDINGS

The *precautionary landing* is a preventive measure. In the mountains, weather changes rapidly. You may find the weather in front closing in just as fast as the weather behind. Disorientation, low fuel, or rough engine may dictate a precautionary landing.

The *forced landing* calls for immediate action by the pilot. The difference between a precautionary and a forced landing is the amount of time the pilot has to react to the situation and, perhaps, the number and quality of possible landing sites. Engine failures, blown oil lines, in-flight fires, and some electrical problems usually result in a forced landing.

Because most training is done in rented aircraft, pilots are taught to pick emergency landing sites that will render the least amount of damage to the aircraft. In a true emergency it may be more advantageous to land in a large rough area rather than a shorter, more smooth one. The best landing site is the one which is most likely to ensure survivability of the occupants and which allows the pilot enough time to plan, control, and execute a safe emergency landing. Survivability of the aircraft is secondary. If a small area is your only choice, plan your touchdown in the middle of the area, even though the aircraft may end up in the trees at the end of the strip. It would be better to hit those trees at a slower airspeed under control and on the ground, than to misjudge the approach and hit the trees on final, eventually hitting the ground out of control and at a much faster airspeed.

Are you a tree top flyer? If so, the sky above you is useless unless when the engine quits. The more altitude you have, the more time you will have for selecting a landing site, declaring an emergency, and preparing your passengers for the emergency landing.

(The following guidelines are for single-engine aircraft emergencies. Multiengine procedures may differ.)

If the engine fails during the takeoff roll, try to remain on the runway. If time permits, go through your emergency shutdown procedures as quickly as possible.

Should your engine fail immediately after takeoff, you must make a decision whether to land straight ahead or, if you have sufficient altitude, turn back and land on the runway. The latter choice is nearly always the most dangerous, yet the one most pilots choose in an emergency. A good rule to help you decide is the "500-foot rule." Most pilots find it difficult to turn their aircraft around 180°, and maintain sufficient airspeed without stalling the airplane unless they have 500+ feet. You may be the exception. Some mountain airstrips are in such rough terrain that you may not have a choice at all. Generally, the rule of thumb is:

> Below 500 feet: Plan to land straight ahead.
> Above 500 feet: Consider returning to the field.

Engine Failure in Flight

If the engine still won't start, *re*check as many things as you can possibly think of, and try to identify the problem. Your emergency checklist will cover most of the possibilities. Magneto switch—maybe you bumped it with your knee during that last jolt of turbulence. Fuel pressure and fuel flow—try flipping on the auxiliary fuel pump or priming the engine with the manual primer. If the fuel line is blocked with ice or water, shooting a primer-full of raw gas into the cylinders may help. You may actually be able to get the engine going—and keep it going for a while—by constantly pumping the primer.

If the engine cannot be restarted, set up for an emergency landing. Altitude and terrain permitting, attempt a normal landing pattern—it is something familiar and will tend to relax you. Touch down slightly tail-low, and apply brakes generously.

Except during periods of high water, sandbars and gravel bars along river routes offer the best places to make emergency landings in remote wilderness areas. Also, highways, railroads, airports, towns, villages, and other settlements tend to be located along rivers and streams.

The Controlled Crash

Always stay in control of the aircraft during touchdown. This means that the airspeed during initial impact should be slightly above stall speed.

If faced with an emergency landing in a forest or trees, locate the largest opening you can find—an area that will allow the fuselage to go through, but sacrificing the wings. Attempting to full-stall the airplane over the trees is not recommended. Remember, keep the airplane under control upon initial impact. If the treetops are high, stalling the aircraft may cause a wing to drop, which would result in the side of the fuselage, or the nose, hitting the ground first.

DON'T

⚠ Don't choose a wet site unless you have absolutely no choice.

⚠ Don't choose sites that are far away from the things you need to survive while you are awaiting rescue. When considering forced landing sites in the mountains, *survival* should be your first concern.

The most common causes of engine-out accidents are indecision, poor airspeed control, poor planning, poor coordination, and poor choice of landing site. Forced landings of any type nearly always benefit from having altitude and holding it as long as possible. Avoid violent maneuvers. Your command of the aircraft is only enhanced by smooth deliberate maneuvers.

FIRE AND ICE—TWO MORE VILLAINS

Whether on the ground or in the air, fire calls for immediate action. For an engine fire, study this checklist and the one in your operator's manual:

Mixture: IDLE CUT-OFF
Fuel Selector or Valve: OFF
Magnetos: OFF
Master Switch: OFF

If fire occurs on the ground, it may be extinguished with a standard fire extinguisher, a wool blanket, or even dirt. If fire occurs in flight, turn off *mixture, fuel selector, mags,* and *master*, and close cabin heat and air vents. Often, engine fires can be extinguished by increasing your glide speed. But doing the same with an electrical fire or a cabin fire may fuel the fire rather than extinguish it. Without panicking, land at the earliest opportunity.

Icing can be a real devil. Mountains, especially, have micro-weather systems that produce icing at altitudes different from those reported by your local FSS.

Careful preflight inspection of control surfaces and propeller can prevent accidents. Aircraft surfaces give up heat faster than the earth. Even in the absence of visible precipitation, if there is enough humidity in the air and the temperature drops below freezing during the night, an aircraft parked outside could end up coated with frost. It must be removed before takeoff.

In flight, ice is more often encountered during IFR conditions than during VFR. However, a word of caution: Reciprocating engines tend to be very susceptible to icing in clear air, even if the ambient air temperature is well above freezing. Moist air cools as it passes through the throat of the carburetor, forming ice there. If the engine begins to run rough:

1. Turn the carburetor heat to ON.

2. Turn the pitot heat switch to ON (as a precaution).

3. Turn back or change altitude.

SURVIVAL IN THE MOUNTAIN WILDERNESS

In the mountains, surviving *after* a forced landing can present a challenge more difficult than surviving the landing itself. You must be prepared with survival gear appropriate to individual needs, temperature, season, and intended routes. Immediately after a forced landing your primary concerns should be:

1. Evacuation from the aircraft until the potential for fire has passed

2. Clothing

3. First-aid for life-threatening injuries

4. Activation of your ELT system

5. Shelter, water, fire, and food

Clothing

Survival clothing should be worn when possible, or kept handy so it will not be lost in the event of fire. In winter, your first line of defense against the elements is what you are wearing when you go down. The warm clothes packed away in your suitcase may not be accessible. You may be too injured to move about or to change clothing. It may be too dark, or raining or snowing such that searching for your luggage, wood, or tools would only worsen your condition and your chance for survival.

The effectiveness of clothing is measured by its ability to trap the body's radiated heat and repel moisture. Wet clothing, for example, will extract body heat 200 times faster than dry clothing. Wool is unique in its ability to maintain body heat fairly well even after it has been immersed in water. Wool traps air in tiny pockets of the fibers, preventing complete saturation. In contrast, cotton saturates very quickly. Cotton also wicks water rapidly from wet places to dry places. Because wool has a relatively loose weave, it is not, however, the best protection against strong winds. A lightweight nylon windbreaker worn over a wool garment will maximize wool's efficiency.

Don't depend on only a single heavy garment. If it becomes soaked you are in trouble. Easily removable *layers* of clothing allow you to better control your own comfort and give some protection against having all of your clothes soaked at once.

New synthetics, such as fiberfill, hollowfill, and polypropylene fabrics, are excellent insulators and may be worn either as outerwear or thermal underwear. They also are resistant to water and are lightweight.

First Aid

Nearly all aircraft crash and survival situations have injuries associated with them, even if only minor ones. First aid should be given promptly to both pilot and passengers. Your survival kit should include at least the first aid items listed in TABLE 7-1.

Table 7-1. *First Aid Kit*

ITEM	QUANTITY	USE
Aspirin	1 Bottle/30 Tablets	Pain
Antacid	6 Tablets Per Person	Indigestion
Antihistamine	6 Tablets Per Person	Bites and Colds
Antiseptic	1 Bottle	Cuts
Adhesive Tape	1 Roll	Bandages
Band-Aids	3 One-Inch Per Person	Lacerations
Butterfly Band-Aids	3 Per Person	Small Cuts
Elastic Bandage	1 Three-Inch Roll	Sprains
Gauze	2 Three-Inch Rolls	Securing Bandages
Eye Dressing	1 Package	Eye Injuries
Salt Tablets	1 Pack of 24	Burn Victims
Burn Ointment	1 Tube	Burns
Scissors	1 Pair	Cutting
Tweezers	1 Pair	Plucking

ELTs

If you crash, stay near the airplane. Most successful rescues have been made when downed pilots and passengers remained with the aircraft. Get away from the immediate vicinity of the plane until the engine has had time to cool down and any leaking fuel has had time to evaporate.

Emergency locator transmitters (ELTs) are automatic radio transmitters which help rescue teams locate downed aircraft. They are required by FAR 91.52 to be on most civilian aircraft. (Canada requires all aircraft overflying its territory to have an ELT on board.) Type F transmitters will normally activate upon impact, however, to ensure that your transmitter is functioning properly, you should manually switch it on. If you are able to land the aircraft without a major impact you must switch on the ELT manually. The transmitters are battery-operated and emit a distinctive audio tone on 121.5 MHz and 243.0 MHz. ELTs will operate continuously for at least 48 hours over a wide temperature range. Most rescues occur within 24 hours after search-and-rescue efforts begin.

Since ELTs operate in the VHF and UHF frequency range they are affected by line-of-sight transmission and reception. To obtain the best range, the transmission should be placed on level ground, as high as possible, and away from obstructions. The antenna should be vertical for optimum radiation of the signal. Placing the transmitter on top of the wing, if it is level, will provide the reflectivity to extend transmission range. Holding the transmitter close to the body in cold weather will not significantly increase battery power output, and could actually reduce the effective range of the transmission.

Remember, high-altitude jets, military aircraft, and many commercial operators routinely monitor the emergency frequencies. Additionally, satellites which operate 20,000 feet above the Earth's surface are also tuned to the emergency frequencies. If your ELT is turned on, it is very likely that the signal will be picked up quickly and the appropriate search-and-rescue team notified.

Ground-to-Air Visual Signaling. If your aircraft is painted with a bright conspicuous color, you are more likely to attract attention from the air. Some pilots carry brilliant colored cloth in their survival kits to attract the attention of low-flying aircraft. Research has shown that royal blue (not red) is the best color for signaling from winter sites. Movement enhances visibility. A lightweight piece of plastic or cloth stretched along a clothesline, flapping in the breeze, helps searchers see you.

Anything that will aid search-and-rescue teams to spot you and distinguish your campsite from natural features on the ground will help. The international signal for distress is "SOS." The SOS signal should be spelled out in large letters in a cleared area visible from all directions to passing aircraft. If you do not have bright-colored fabric or plastic for making the letters, use pieces of wood, stone, or anything handy that will contrast with the background.

The international distress signal using campfires is three signal fires placed in a triangle roughly 25 feet apart. A rule of thumb when signaling by fires is "smoke by day" and "fire by night." Green boughs or grass on a hot fire produces white smoke. Rubber, plastic, or oil produces black smoke.

Shelter

The aircraft itself is an excellent source of survival materials. After all danger of fire has passed, the cabin can make a superb overnight shelter. Since metal, especially aluminum, conducts heat rapidly, when night comes, temperature inside the cabin will dissipate rather fast. You should make an effort to insulate the metal walls and close off any openings. Use newspapers, charts, cartons, and fabrics to insulate and curtain off all the interior space that you do not actually use. This will reduce the drain on your own body heat. Stuff cracks and crevices with any suitable material you can find to seal yourself tightly inside. The outside air temperature will generally fall steadily throughout the night. The colder you get, the less able you are to function.

If the cabin is damaged too badly or access into the plane is difficult, you may choose to use parts of the aircraft to form a windbreak and shelter from the outside elements. Wings, tail section, carpeting, pieces of cowling, tree limbs—even blocks of snow—work well in a pinch. If you must construct a shelter on the ground, insulate the floor. Use any nonconductive material to keep your body away from the damp, cold earth: seat cushions, carpeting, sunvisors, plastic interior side panels, or floorboards.

TABLE 7-2 lists some uses you may find for the various parts of your aircraft.

Survival Techniques

Mountain flights should begin with common sense. Before leaving, whether on a short day trip, an overnight outing, or a full-fledged vacation, pilots should file a flight plan or at a very bare minimum leave a "trip plan" at home with someone who would be in a position to initiate search-and-rescue procedures in the event you are overdue. Trip plans should include at least the following information:

- ⚠ Names of pilot and passengers
- ⚠ Description of the aircraft
- ⚠ Purpose of the trip
- ⚠ Departure time
- ⚠ Intended route
- ⚠ Destination
- ⚠ Estimated time of arrival

163

Table 7-2. *Emergency Uses of Aircraft Parts*

Ailerons—snow-cutting tools; shelter braces; splints
Air Filter—fire-starter material; water filter
Aluminum Skin—reflector for warmth around a fire; splint; signaling device; snow saw blade
Battery—signaling with lights; fire starter
Brake Fluid—fire starter
Charts and Maps—stuff inside clothing for insulation
Compass—oil for starting fire; direction finding
Control Cables—rope; snare wire; binding for shelter
Control Pulleys with Cable—block and tackle
Disc Brake Plates—signaling devices
Doors—shelter; windbreak
Engine Cowl—shelter; water collector; windbreak; fire platform
Engine Mags—spark producers for starting fires
Engine Oil—fire starter; black smoke for signaling
Engine Gas—fire starter; fuel for stove; signaling
Fabric Skin—fire starter; water collector
Fuel Cells—use to melt snow on a black surface; black smoke for signaling; place on snow for signal to
    search-and-rescue planes
Fuselage—shelter
Hoses—siphon for water/gas/oil; burn for black smoke
Inner Tubes—canteen; elastic binding; black smoke signal
Interior Fabric—water strainer or filter; clothing; insulation; bandages
Landing Light Lens—fire starter
Landing Light, Strobes—use with battery to signal at night
Light Covers—utensils; small tools
Magnesium Wheels—signaling devices
Nose Spinner Cone—bucket; stove with sand, oil, and fuel; scooping tool; cooking pot; funnel
Oil Filter—burn for black smoke
Propeller—shovel; snow-cutting tool; shelter brace
Rotating Beacon Lens—drinking cup
Rugs—ground pad; insulation, clothing
Seats—sleeping cushions; back brace; fire starter; signal material; insulation; ground pad; rubber
    sponge for neck support
Seatbelts—binding material; slings; bandages
Spring Steel Landing Gear—pry bar; splint
Tires—fire starter; fuel; black smoke
Vertical Stabilizer—shelter support; platform
Wheel Fairing—water storage; water collection; black smoke
Windows—shelter; windbreaks; cutting tools
Wings—windbreaks; shelter supports; overhead shade; platform for fire; water collector; signaling device; crutch
Wingtips—drip collectors; water carriers
Wiring—binding, rope
Wooden Wing Struts, Braces, Props—fire starter; fuel; pry bar; splint; shelter brace; flag pole

Table 7-3. *Mandatory Survival Equipment for Canada and Alaska*

ALASKA

Alaska law requires the following emergency equipment for all flights within the state:

- ☐ Food for each occupant to sustain life for two weeks
- ☐ One axe or hatchet
- ☐ One first-aid kit
- ☐ One pistol, revolver, shotgun, or rifle, and ammunition for same
- ☐ One small gill net and an assortment of tackle, such as hooks, flies, lines, sinkers, etc.
- ☐ One knife
- ☐ Two small boxes of matches
- ☐ One mosquito headnet for each occupant
- ☐ Two small signaling devices, such as colored smoke bombs, railroad fuses, or Very pistol shells in a sealed metal container
- ☐ One pair of snowshoes (*October 15 to April 1 only*)
- ☐ One wool blanket for each occupant over four years of age (*October 15 to April 1 only*)

Note: Requirements vary for large multiengine aircraft.

CANADA

Flight plans are required for all Canadian flights.
Canadian law requires the following emergency equipment:

- ☐ Food having a caloric value of at least 10,000 calories per person carried, not subject to deterioration by heat or cold and stored in a sealed waterproof container bearing a tag or label on which the operator of the aircraft or his representative has certified the amount and satisfactory condition of the food in the container following an inspection made not more than six months prior to the flight
- ☐ Cooing utensils
- ☐ Matches in a waterproof container
- ☐ Stove and a supply of fuel or a self-contained means of providing heat for cooking when operating north of the tree line
- ☐ Portable compass
- ☐ Axe of at least 2½ lbs. or 1 kg weight with a handle of not less than 28 in. or 70 cm. in length
- ☐ Flexible saw blade or equivalent cutting tool
- ☐ Snare wire of at least 30 ft. or 9m and instructions for its use
- ☐ Fishing equipment including still fishing bait and a gill net of not more than a 2 in. or 5cm mesh
- ☐ Mosquito nets or netting and insect repellant sufficient to meet the needs of all persons carried when operating in an area where insects are likely to be hazardous
- ☐ Tents or engine and wing covers of suitable design and color, or having panels colored in international orange or other high visibility color, sufficient to accommodate all persons carried when operating north of the tree line
- ☐ Winter sleeping bags sufficient in quantity to accommodate all persons carried when operating in an area where the mean daily temperature is likely to be 7 °C or less
- ☐ Two pairs of snow shoes when operating in areas where the ground cover is likely to be 12 in. or 30cm or more
- ☐ Signaling mirror
- ☐ Three or more pyrotechnical distress signals
- ☐ Sharp jackknife or hunting knife of good quality
- ☐ Suitable survival instruction manual
- ☐ Conspicuity panel

Suggested additional items:

- ☐ Spare axe handle
- ☐ Honing stone or file
- ☐ Ice chisel
- ☐ Snow knife or snow saw-knife
- ☐ Snow shovel
- ☐ Flashlight with spare bulbs and batteries
- ☐ Pack sack

Note: In Canada, firearms are carried at the operator's discretion. Small arms (handheld pistols, revolvers, etc.) and fully automatic weapons are prohibited. Firearms must be declared to Canadian Customs.

⚠ Expected time of return (if round trip)

⚠ List of alternate destinations

⚠ List of emergency equipment carried on board

If you make a regular practice of flying in the mountains or sparsely populated areas, invest in a survival kit and carry it with you in your aircraft at all times. Survival kits may be purchased from commercial suppliers or put together yourself. Flights conducted through Canada and Alaska are required by law to carry the survival items listed in TABLE 7-3.

Building a Fire. While awaiting rescue, you will need fire for warmth, for keeping dry, for signaling, for cooking, and for purifying water. Fires should be kept small. Large fires require too much fuel and are not as easy to control. In cold weather, several small fires arranged in a circle around a person are much more effective than one large fire.

Fire building is a three-stage process: *tinder*, *kindling*, and *fuel*. Tinder is very fine, highly flammable, dry material which takes a spark to a flame or picks up a flame from a match. Kindling is small, highly flammable, dry material which is larger than tinder and will take the flame from the tinder and make it hot enough and big enough to ignite fuel. The following are sources of tinder, kindling, and fuel:

⚠ *Tinder*: fine dry twigs, fine shavings, dry bark, dry moss, dry leaves, loose ground-lying lichens, ferns, dead grass, straw, or other lightweight material. Birch bark, pitch, pinecones, charcoal lighter fluid, gas, and oil will serve the purpose when conditions are wet or moist.

⚠ *Kindling*: sticks, larger twigs, crumpled paper, empty cardboard boxes, paper bags. Please note that fabrics from the cabin of your aircraft are fire-retardant materials and consequently are not good as kindling or fuel.

⚠ *Fuel*: larger sticks, branches, dry livestock dung, twisted dead grass stalks, dry logs. Note that liquid fuels, like gasoline and oil, are quickly exhausted and will not maintain a fire for any length of time.

In the arctic, low, dead, needle-bearing branches of standing spruce trees are excellent fuel. On the tundra, wood is scarce. Collect low brush or shrub material or roots. Dry grasses are usually plentiful. Along the coast look for driftwood. Animal fat can be used as fuel by putting chunks of fat on a stick-or-bone framework. Mosquitoes menace the arctic and tundra areas of the North. Burning green material will make a smudge that will keep them away.

In the desert, fuel may be extremely scarce. Dry animal dung, found along commonly traveled routes, will give off a very hot flame. If there is plant growth of any type, remember that plants have roots, leaves, and twigs which make good fuel. Deserts cool considerably after the sun goes down. Don't disregard the need for a fire just because you were sweltering shortly before sunset.

Choose the location of your fire carefully. To get the most warmth from your fire, and to protect it from the wind, build the fire close to or against a large rock. In areas without rocks, construct a wall out of logs or a section of the aircraft (wing, elevator, vertical stabilizer) in order to reflect and direct the heat into your shelter. Do not build a fire inside any part of the aircraft, under a wing, or any place which might ignite residual fuel. Scrape down to bare dirt, or on ice and snow, build a platform out of logs, stones, or a piece of metal from the aircraft.

Fire climbs upward. Always add kindling and fuel from above the flame. To make the fire last all night long, place a large log over it so that the fire will burn into the heart of the log. When a good bed of coals has formed, cover the bed lightly with ashes first then dry earth. In the morning the fire will still be smoldering.

You can *start a fire without matches* by any of the following means:

- ⚠ *Flint and Steel*—Scrape a knife blade against a very hard rock or the bottom of a waterproof matchbox to produce a spark.

- ⚠ *Lens or Mirror*—A mirror or any convex lens from a camera, binoculars, etc., can be used in bright sunlight to focus the sun's rays on tinder and start it burning.

- ⚠ *Aircraft Battery*—Scratch the ends of battery cables or wires together to produce an arc. Use the spark to ignite a fuel-soaked rag.

- ⚠ *Cigarette Lighter*—Even if out of fuel, a cigarette lighter is a good source of sparks as long as the flint will last.

Short-Use Fires. For cooking, signaling, or boiling water when wood is not readily available, improvise with a metal can. Partially fill the can with sand or dirt. The can should have breather holes—around the rim for smoke, and at sand level for oxygen. Pour one-half cup of avgas on the sand in the can. Wait two minutes, then toss a match in the can. The initial flare-up will settle down to a steady, safe flame. Oil mixed with gasoline will produce a lower, longer-burning flame. If you don't have a metal container, then gasoline, or oil and gas mixed, can be poured in a hole or directly onto the ground to achieve the same effect.

Water

One of the most important needs when stranded in the wilderness is water. Remember, most rescues are completed in 24–48 hours, but they can and do take longer. Alaska and Canada require, for each occupant on board, survival provisions to sustain them for at least two weeks. It may take searchers two weeks or longer to locate you, due to weather and/or the sheer vastness of the search area. A person can survive for weeks without food, but most people will survive only five days or less without water.

Normally a person needs two quarts of water per day to maintain efficiency. Smaller intakes result in a loss of efficiency. If you delay drinking, your body must make it up later on. Dehydration can be just as serious a problem in cold areas as it is in the desert.

Purify all water before you drink it. This applies to wilderness areas especially. That pristine little brook cascading over and around boulders and stones is literally filled with microorganisms that can cause severe diarrhea and, subsequently, dehydration. Either boil the water or use purification tablets. Rainwater collected from plants or clean containers is generally safe to drink without purification.

Ice is better to melt than snow because you get more water for the amount of fuel you must burn to melt it. In rocky country, look for springs and seepages. Limestone country seems to have more springs than other types of rocky terrain. Cold springs are less likely to be polluted than warm or hot springs. Seepages are found where a dry canyon cuts through a layer of porous sandstone or where "the grass is greener." Water is more abundant and easier to find in loose sediments than in rocks. Flat benches or terraces above river valleys frequently have springs and seepholes along the base, even when the river bed is dry.

In stony desert country, dry stream beds may be your best source of water. Dig a hole at the lowest point on the outside bend in the stream channel. Often the sun has dried only the upper few inches of the sand, and water or damp sand lies just below the surface. Desert country is a great place for collecting dewdrops. It is possible to collect as much as one quart per hour from a dew collector system. Dew forms on the underside of metal surfaces, rocks, and sometimes logs. Mop up the dew with a piece of cloth and squeeze it into a container.

Food

Take careful stock of your available food and water. Estimate the number of days you think you will be on your own and then divide your food supply into thirds. Allow two-thirds for the first half of your estimated time before rescue, and save the remaining one-third for the second half. Your energy requirements will be higher at first while you are still recovering (possibly from mild shock) from the forced landing and seeing to immediate needs of warmth, shelter, signal fires, water, and food. Every bit of work requires additional food and water.

The less active you are, the less food and water you will need.

If you have less than a quart of water per day in reserve, avoid dry, starchy, and highly flavored foods and meat. Eating increases thirst. Where water is in short supply, the best foods to eat are those high in carbohydrate content. Carbohydrates are foods which come primarily from plants, such as sugar, starches, cereals, and fruits.

The best sources of protein are foods which come from animals, such as meat, fish, eggs, milk, and cheese. Proteins are valuable fuels and are responsible for maintaining and repairing body tissues. Although you can subsist without proteins, normal body intake should average about three ounces per day. If your water supply is limited, avoid eating large amounts of protein.

Fats, such as olive and cottonseed oil, vegetable oil, lard, and butter, come from either plant or animal. They are not generally necessary for human nutrition and they require added water for elimination.

Prepared Foods. The foods in most survival kits have been developed especially to provide proper sustenance in emergencies. Eating these foods as directed on the package will keep you at maximum efficiency. Ideally, you should try cooking and eating survival kit foods at home first so that in a real emergency there will be no surprises.

Eating Off the Land. If you crash in a remote wilderness area, like Alaska, Canada, or many parts of the Rocky Mountains, rescue may be slow in coming. Plan to supplement your survival rations with native foods.

There are many wild foods that are rich in vitamin and mineral content. Fleshy-leafed plants make good green salads or vegetables. In Alaska we have the fiddle-head fern. It tastes a lot like spinach; it is delicious and very nutritious. Other parts of the country have their specialties too. Fresh fruits and berries are an excellent supplement to a restricted water supply. Nearly all animals are edible when freshly killed. Use snare wire from your survival kit to set snares along game trails. You can eat any kind of bird or bird eggs, frogs, turtles, lizards, and all snakes except sea snakes. Do not eat toads. If you don't know the difference between a frog and a toad, avoid eating them altogether. Game is easiest to find near water, in clearings, and along the fringes of thickly forested land.

If you have a gun or homemade spear, "ambushing" your dinner is probably your best bet. By hiding downwind and along well-traveled trails, near water, or near feeding grounds, you are more apt to be successful. Animals move to feed and water in the early mornings or late evenings. In your travels near your campsite, stay alert for signs such as tracks, trampled underbrush, and droppings. On narrow trails, be ready for game using the same path.

Fishing is a terrific source of fresh food. The most common way to fish is with a line and hook, which should be in your survival kit. Use insects, shellfish, worms, or pieces of meat for bait. Artificial lures can be made from patches of brightly colored cloth. If you do not have a line and hook you can improvise using aircraft cable or wire, and pieces of metal for lures. The most efficient way to fish is with a net. Some survival kits include a "gill net." Nets should be placed in quiet water, anchored with stones and supported with floating chunks of wood.

Eat regularly. Don't nibble. With limited rations, plan to eat at least one good meal per day. Sit down and make a real feast of it. Two meals a day are preferred. One meal should be served and eaten hot, if possible. If you need to collect wild foods to supplement your stocks, cooking makes wild foods safer, more digestible, and generally more palatable. Native foods may be more appetizing when they are eaten separately. Don't mix your survival rations with foods you collect.

IN SUMMARY

Mountain flying almost always takes a pilot over remote, hard-to-get-to places on at least a portion of his flight plan. Search and rescue in these areas has become increasingly sophisticated and dependable thanks to the combined efforts of the FAA, Civil Air Patrol, Coast Guard, Department of Fish & Wildlife, and a whole host of others. With ELTs, transponders, and two-way radios, pilots can let others know that they are in trouble and require assistance. This improved chance of eventual rescue makes it ever more vital for pilots to "be *prepared* to survive."

Appendix A
The Fine Art of Air Drops
by Doug Geeting

SOME TIME AGO I WAS HAVING COFFEE IN THE TALKEETNA ROAD HOUSE, AND the topic of discussion with a couple of local pilots was "air drops"—a nice way to describe throwing things out of airplanes.

I get a lot of calls for air drops, mainly from my work on Denali rescue efforts. Mountain climbers who get caught in a bad storm may need extra supplies. Air drops are the only way to get them what they need.

I have dropped everything from turkeys at Thanksgiving to halibut at 14,000 feet on Denali. I have dropped 2×6×10-foot pieces of lumber for remote cabin builders and oxygen bottles to a pulmonary edema victim at 19,200 feet. Most of my air drops have been completed with a reasonable measure of accuracy, but there have been some problems encountered by me in the airplane, and some by the recipients on the ground!

A couple of years ago a high-altitude medical camp was set up at the 14,300-foot level on Denali. The staff needed to have fresh supplies dropped about every two weeks, along with their mail and occasionally some beer and ice cream.

On one of these flights, I was instructed to drop a case of beer, climbing rope, two small boxes of dried food, and one 75-pound halibut. I cut the halibut into two long pieces so it could fit through the camera hatch in the belly of my Cessna 185. I then loaded everything into the aircraft, putting the items in last that I wanted to drop first. I always load the plane this way to avoid confusion later.

Since I had quite a bit of gear and the two large hunks of fish, I stopped at base camp on the Kahiltna glacier at 7,300 feet and picked up an "assistant air dropper" to help me with the load. Having someone else drop for me would make things go smoothly, or so I thought.

We took off from base camp and headed for the 14,300-foot camp. I instructed my lovely "air dropping assistant" to drop the objects when I told her to. We came around and lined up for the target zone, which was in line with and about 500 feet away from the camp. Susan asked me what to drop first.

I yelled back, "Halibut!" It looked like a perfect drop.

As I turned to get lined up for the next drop, I looked around to check on Susan and the gear. She was a little queasy from the bumping around and the steep turns, and then I noticed—there was nothing left in the airplane!

"What happened to all the cargo?" I shouted.

"You said to drop 'all of it'!" she replied, indignantly.

"I said, 'halibut', not 'all of it'!"

I turned to fly over camp to see where the beer and the food boxes had ended up. Everyone in camp was running around frantically, and when they saw me coming at them again they actually began taking cover!

The "mad bombardier" of Talkeetna had devastated two dome tents, one with the halibut and the other with the beer. A 40-pound slab of fish hit the side of one of the tents and landed right between the two occupants inside. It was lucky the stray fish did not hit one of them, as the tent was literally torn away from underneath them!

The beer hit another tent and exploded on impact. The two occupants must have thought the gods were raining beer upon them. I certainly hope they didn't have nightmares after that experience—I know *I* did!

I straightened things out with the camp, but I learned one very important lesson: good communication is important when doing air drops. In most cases, you are working low to the ground and slower than usual. Distractions of any kind are to be avoided.

After that I had an intercom system installed in the 185 for all six seats. With this excellent communication, my helper can hear my instructions clearly, I can fly the airplane, and the air drop is made at the right time and in the right order.

Another important thing to remember: doing an air drop by yourself can mean trouble. A very good pilot and friend of mine in Talkeetna attempted to perform an air drop across the Susitna River near town. This fellow is very large with broad shoulders and flies a Cessna 170. The pilot slowed the plane and stuck the object to be dropped out the window on his side.

In the process, he also put his shoulders out, got them stuck and could not get back inside. *Crash!* He hit the trees, ruined his airplane, but fortunately only received a broken lip!

If I have to make a drop by myself, I always drop the object out of the window on the passenger side. That way I cannot get my shoulders out and am in a better position to grab the throttle or control yoke.

Accuracy in air drops is, of course, very important, and the only way to become accurate is to practice. It just takes a few times to get the hang of it. In my 185, I slow down to about 100 MPH (IAS) with two or three notches of flaps, and get the plane trimmed so as not to fight the controls. Then I line up with my target at about one-half mile or farther away, to allow time for getting in proper alignment with the target.

In a side-by-side aircraft, such as a Cessna, I line up the target with the row of rivets along the cowling. Then, I move the aircraft to the left very slightly, so that the target is placed off to the right, almost between the row of rivets and the nose of the aircraft. I fly at an altitude of about 500 feet or slightly lower if there are no trees in the way.

The object is dropped at about the time the target disappears from view under the nose. This method is a good one for any pilot to start with. The altitude and drop point can be varied as your skills improve.

All objects I drop are marked with red or orange survey tape. This is easy to see from the ground, and even if the objects are dropped into deep snow and disappear from view, the red survey tape will remain visible. When dropping in water or swampy areas, tie clorox bottles (with the caps tightly fastened) to the objects and then tie survey tape to them. The bottles will float and the tape will make it easier to locate the objects.

One important thing to remember: gather the tape or rope tightly to the side of the object to be dropped; this will help keep anything from getting tangled up on your elevator. If I drop out of the right door or window, I will yaw the airplane's nose to the right just before I drop, again to ensure that nothing gets wrapped around the elevator. It is important to keep up your speed while dropping, to allow you to yaw the plane considerably without danger of stalling out your inside wing. This will also allow a little time in case you have to get to the throttle and controls to increase your speed.

FAR Part 91.13, *Dropping Objects*, says, "No pilot in command of a civil aircraft may allow any object to be dropped from that aircraft in flight that creates a hazard to persons or property. However, this section does not prohibit the dropping of any object if reasonable precautions are taken to avoid injury or damage to persons or property."

Air drops can be performed safely, as long as certain precautions are taken. Just remember, pay attention to what you are doing, use good communications with your assistant, and don't allow yourself to be distracted.

Appendix B
Helpful Formulas

HERE ARE SOME RELATIONSHIPS THAT YOU SHOULD KNOW:

⚠ Multiply MPH by 0.868 to get knots

⚠ Multiply knots by 1.152 to get MPH

⚠ Multiply MPH by 1.467 to get ft/sec

⚠ Multiply knots by 1.69 to get ft/sec

⚠ The formula for radius of turn when your airspeed is in knots is:

$$R = \frac{V^2}{11.27 \tan \phi}$$

where: V = velocity (in knots)
R = radius of turn (in feet)
ϕ = bank angle

The formula for radius of turn when your airspeed is in MPH is:

$$R = \frac{V^2}{14.96 \tan \phi}$$

where: V = velocity (in MPH)
R = radius of turn (in feet)
ϕ = bank angle

ϕ	20°	30°	40°	45°	50°	60°	70°	75°
$\tan \phi$	0.36	0.58	0.84	1.0	1.19	1.73	2.75	3.73

▲ If you want to know the normal acceleration (g-forces) that you will pull in a level turn at any bank angle, use the following relationship (it holds for any airplane):

$$N_z = \frac{1}{\cos \phi}$$

where: N_z = normal acceleration (in g's)

ϕ = bank angle

▲ If you want to know how much the stall speed increases when you are pulling g's, use the following:

$$V_g = V_{stall} \sqrt{N_z}$$

where: V_g = accelerated stall speed

V_{stall} = 1-g stall speed

N_z = the g's you are pulling

ϕ	20°	30°	40°	45°	50°	60°	70°	75°
$\cos \phi$	0.94	0.866	0.77	0.707	0.64	0.5	0.34	0.26

Appendix C
Manufacturers and Suppliers

Aircraft Manufacturers

Aerospatiale General Aviation
1900 Westridge
Irving, TX 75038
(214) 550-7433

Alexander Aeroplane Co.
900 S. Pinehill Rd.
Griffin, GA 30123
(404) 228-3815
(800) 831-2949

Beech Aircraft Corp.
9709 E. Central
Wichita, KS 67201
(316) 681-7111
(800) 835-7767

Bellanca
Miller Flying Service
P.O. Box 190
Plainview, TX 79072
(806) 293-4121

Cessna Aircraft Co.
P.O. Box 7706
Wichita, KS 67201
(316) 946-6056

DeHavilland Aircraft, Ltd.
Downsview
Ontario, Canada M3K 1Y5

Gulfstream Aerospace Corp.
P.O. Box 2206
Savannah, GA 31402
(912) 964-3233

Helio Aircraft, Ltd.
P.O. Box 604
Pittsburg, KS 66762
(316) 231-0200

Lake Aircraft
Laconia Airport—Hangar #1
Laconia, NH 03246
(603) 524-5868

Maule Aircraft
EDO Corp.
Lake Maule, Rte 5, Box 319
Moultrie, GA 31768
(912) 985-2045
(718) 445-6000

Mitsubishi Aircraft Int'l.
P.O. Box 3848
San Angelo, TX 76901
(214) 387-5600

Mooney Aircraft Corp.
P.O. Box 72
Kerrville, TX 78028
(512) 896-6000
(800) 356-6931

Piper Aircraft Corp.
P.O. Box 1328
Vero Beach, FL 32960
(305) 567-4361
(800) 727-4737

Rockwell International
General Aviation Div.
5001 N. Rockwell Dr.
Bethany, OK 73008

Taylorcraft Aviation Corp.
P.O. Box 2625
Alliance, OH 44601
(216) 823-6675

Conversion Specialists

Aero Mods
Paine Field, Bldg. #C-3
Everett, WA 98204

Aircraft Conversion Technologies
1410 Flight Line Dr., Hangar A
Lincoln Airport, CA 95648

Ayres Corp.
P.O. Box 3090
Albany, GA 31706

Colemill Enterprises
P.O. Box 60627
Nashville, TN 37206

Flint Aero, Inc.
8665 Mission George Rd. #D-1
Santee, CA 92071

Horton, Inc.
Wellington Municipal Airport
Wellington, KS 67152

Met-Co-Aire
P.O. Box 2216
Fullerton, CA 92633
(714) 870-4610

Riley Aircraft Corp.
2016 Palomar Airport Rd.
Carlsbad, CA 92008

Robertson Aircraft Corp.
839 W. Perimeter Rd.
Renton Municipal Airport
Renton, WA 98666

Sierra Industries, Inc.
P.O. Box 5184
Uvalde, TX 78802

Seguin Aviation
2075 Highway 46
Seguin, TX 78155

Turbotech
P.O. Box 61586
Vancouver, WA 98666

Emergency and Safety Equipment (including ELTs)

Airborne
1160 Center Rd.
Avon, OH 44011

ARP Industries
36 Bay Dr.
East Huntington, NY 11743

Autotec, Inc.
P.O. Box 391
Sylvania, OH 43560

BRS, Inc.
1845 Henry Av.
So. St. Paul, MN 55075
(612) 457-7491

Dale & Associates
1401 Cranston Rd.
Beloit, WI 53511

DeVore Aviation Corp.
16160 Stagg St.
Van Nuys, CA 91406

DME Corp.
1631 S. Old Dixie Hwy., Bldg. E
Pompano Beach, FL 33060

Emergency Beacon Corp.
13 River St.
New Rochelle, NY 10801

Intertech Aviation Services
3 Sunset Ln.
Littleton, CO 80121
(303) 781-4177

LA Screw Products
Police Equipment Division
8401 Loch Lomond Dr.
Pico Rivera, CA 90660

Martech
P.O. Box 1539
Ft. Lauderdale, FL 33302

Tanis Aircraft Services
P.O. Box 117
Glenwood, MN 56334

Wilbur Industries
Southwest Harbor, Maine 04679

Flight Instruments and Navigation Equipment

Aerosonic
3312 Wiley Post Rd.
Carrollton, TX 75006

Artais Weather Check
4660 Kenny Rd.
Columbus, OH 43220

Astronautics Corp.
907 S. First St.
Milwaukee, WI 53204

Arnav Systems, Inc.
Box 23939
Portland, OR 97223
(503) 684-1600
(800) 888-7628

Bendix/King
General Aviation Avionics Div.
400 N. Rogers Rd.
Olathe, KS 66062
(913) 782-0400

Bonzer
90th and Cody
Overland Park, KS 66214

Brittain Industries, Inc.
P.O. Box 51370
Tulsa, OK 74151

Century Instruments Corp.
4440 Southeast Blvd.
Wichita, KS 67210

Davtron, Inc.
427 Hillcrest Way
Redwood City, CA 94062
(415) 369-1188

EDO-AIRE
216 Passaic Ave.
Fairfield, NJ 07006

Foster Airdata Systems, Inc.
7020 Huntley Rd.
Columbus, OH 43229
(614) 888-9502

Global Navigation
2144 Michelson
Irvine, CA 92715

Instruments and Flight Research
2716 George Washington Blvd.
Wichita, KS 67210

II Morrow, Inc.
P.O. Box 13549
Salem, OR 97309
(800) 742-0077
(800) 654-3415 (*Canada*)

Jet Electronic & Technology
5353 52nd St.
Grand Rapids, MI 49508

Micrologic Avionics, Inc.
20801 Dearborn
Chatsworth, CA 91311

Narco Avionics
270 Commerce Dr.
Fort Washington, PA 19034
(215) 643-2900
(800) 223-3636

Nelco
7095 Milford Industrial Rd.
Baltimore, MD 21208

Northstar Avionics
30 Sudbury Rd.
Acton, MA 01720
(617) 897-6600
(800) 628-4487

Offshore Navigation, Inc.
P.O. Box 23504
New Orleans, LA 70183

Rockwell International
Collins General Aviation Div.
400 Collins Rd., N.E.
Cedar Rapids, IA 52406

Safe Flight Instrument Corp.
P.O. Box 550
White Plains, NY 10602

SRD Labs
381 McGlincy Ln.
Campbell, CA 95008

STS Avionic Products, Inc.
11600 Lilburn Park Rd.
St. Louis, MO 63146
(800) 231-5322

Symbolic Displays, Inc.
1762 McGaw Ave.
Irvine, CA 92714

Terra Corp.
3520 Pan American Freeway, N.E.
Albuquerque, NM 87107
(505) 884-2321

Texas Instruments, Inc.
P.O. Box 405, M/S 3438
Lewisville, TX 75067

Tracor Aerospace Group
6500 Tracor Ln.
Austin, TX 78721

Floats

Capre Aqua Floats
805 Geiger Rd.
Zephyrhills, FL 33599

Devore PK Floats
6104B Kircher Blvd., N.E.
Albuquerque, NM 87109
(505) 345-8713

EDO Floats
65 Ruchmore St.
Westbury, NY 11590

Fiberfloat
895 East Gay St.
Bartow, FL 33830

Wipline Floats
Wipaire, Inc.
South End Doane Trail
Inver Grove Heights, MN 55075

Oxygen Systems

Aerox Aviation Oxygen Systems, Inc.
80 Ferry Blvd.
Stratford, CT 06497
(203) 377-5849
(800) 237-6902

The Ted Nelson Co.
3400 San Juan Dr.
Reno, NV 89509
(702) 323-4955

Puritan-Bennett Aero Systems Co.
111 Penn St.
El Segundo, CA 90245
(213) 722-1421

Scott Aviation
225 Erie St.
Lancaster, NY 14086
(716) 683-5100

Sky-Ox, Ltd.
P.O. Box 500
St. Joseph, MI 49085
(616) 925-8931

White Diamond Corp.
P.O. Box 8698
Calabasas, CA 91302
(818) 348-4110

Radar and Radio Equipment

Aircraft Radio & Control
Div. of Cessna Aircraft Co.
P.O. Box 150
Boonton, NJ 07005

Bendix/King
General Aviation Avionics Div.
400 N. Rogers Rd.
Olate, KS 66062
(913) 782-0400

Brellonix, Inc.
106 North 36th St.
Seattle, WA 98106

Comant Industries, Inc.
3021 Airport Ave.
Santa Monica, CA 90425

David Clark Company
Box 155
Worcester, MA 01613
(617) 756-6216

Dayton-Granger
P.O. Box 14070
Ft. Lauderdale, FL 33302

Dorne & Margolin, Inc.
2950 Veterans Memorial Hwy.
Bohemia, NY 11716

General Aviation Electronics
4141 Kingman Dr.
Indianapolis, IN 46226

Bendix/King
400 North Rogers Rd.
Olathe, KS 66062
(913) 782-0400

Martech
P.O. Box 1539
Ft. Lauderdale, FL 33302

Mentor Radio Co.
1561 Lost Nation Rd.
Willoughby, OH 44094

Multitech
8477 Enterprise Way, #101
Oakland, CA

Plantronics
345 Encinal St.
Santa Cruz, CA 95060

Radio Systems Technology
10985 Grass Valley Ave.
Grass Valley, CA 95945

Sigtronics
822 No. Dodsworth Ave.
Covina, CA 91724
(818) 915-1993

Sperry Flight Systems
Avionics Div.
P.O. Box 9028
Van Nuys, CA 91409

Superex Electronics Corp.
151 Ludlow St.
Yonkers, NY 10705

3M Aviation Safety Systems
6530 Singletree Dr.
Columbus, OH 43229
(614) 885-3310

Telex
9600 Aaldrich Ave., S.
Minneapolis, MN 55420

Terra Corp.
3520 Pan American Freeway, N.E.
Albuquerque, NM 87107
(505) 884-2321

Skis

Fluidyne Engineering
5916 Olson Hwy.
Minneapolis, MN 55422
(612) 544-2721

About the Authors

Owner-operator of Doug Geeting Aviation / Mountain Air Transport in Talkeetna, Alaska, Doug Geeting (above) is recognized as one of the most experienced professional mountain/glacier pilots in the United States.

Doug has been a licensed pilot for 18 years and a professional mountain and glacier pilot for 15 of those years. His flying experience, now totaling more than 14,000 hours, has been acquired primarily in the Alaska Mountain Range, which contains the largest and some of the most rugged mountains in North America.

Doug is a certificated flight instructor, advanced ground instructor, and commercial pilot with airplane-single-engine-land and -sea, multiengine, instrument, and glider ratings. A polished airshow performer, he appears regularly throughout Alaska, piloting a Pitts Special and Taylorcraft.

Throughout his career, Doug has concentrated on developing special high-altitude mountain flying techniques. In 1977 he was credited with one of the highest landings ever made on Mt. McKinley, in a single-engine non-turbocharged Cessna 185, when he flew National Park Service officials to the 14,300-ft level to establish a base for high-altitude rescue missions.

Steve Woerner is president of Alaska Pacific Research Company in Anchorage, Alaska. His research and writing experience spans more than 24 years, analyzing methodology and preparing technical documentation for corporate clients. As a free-lance photojournalist, he has published numerous articles in national magazines covering topics ranging from history to flying to politics. He has authored and co-authored three books, including *The Alaska Handbook* (McFarland & Co., 1986). A pilot for more than 12 years, Steve is a certificated flight instructor for single-engine airplanes and holds a commercial pilot certificate with single-engine, multiengine, and instrument ratings.

Index

Edited by Carl H. Silverman
Designed by Jaclyn Saunders and Carl H. Silverman

Other Bestsellers of Related Interest

GOOD TAKEOFFS AND GOOD LANDINGS
2nd Edition—Joe Christy,
revised and updated by Ken George

This second edition is a complete guide to safe, precise takeoffs and landings. This updated edition includes new material on: obstructions to visibility, wind shear avoidance, unlighted night landings, and density altitude. You'll also find a recap of recent takeoff and landing mishaps and how to avoid them, expanded coverage of FARs, and information on the new recreational license. Special emphasis is placed on precision, and on safe practices that should become important habits. 208 pages, 76 illustrations. Book No. 3611, $15.95 paperback, $25.95 hardcover

ABCs OF SAFE FLYING—2nd Edition
—David Frazier

Attitude, basics, and communication are the ABCs David Frazier talks about in this revised second edition of a book that answers all the obvious questions, and reminds you of others that you might forget to ask. This edition includes additional advanced flight maneuvers, and a clear explanation of the Federal Airspace system. 198 pages, 69 illustrations. Book No. 2430, $12.95 paperback only

Prices Subject to Change Without Notice.

Look for These and Other TAB Books at Your Local Bookstore

To Order Call Toll Free 1-800-822-8158
(in PA, AK, and Canada call 717-794-2191)

or write to TAB Books, Blue Ridge Summit, PA 17294-0840.

Title	Product No.	Quantity	Price

☐ Check or money order made payable to TAB Books

Charge my ☐ VISA ☐ MasterCard ☐ American Express

Acct. No. _____ Exp. _____

Signature: _____

Name: _____

Address: _____

City: _____

State: _____ Zip: _____

Subtotal $ _____

Postage and Handling
($3.00 in U.S., $5.00 outside U.S.) $ _____

Add applicable state and local
sales tax $ _____

TOTAL $ _____

TAB Books catalog free with purchase; otherwise send $1.00 in check or money order and receive $1.00 credit on your next purchase.

Orders outside U.S. must pay with international money order in U.S. dollars.

TAB Guarantee: If for any reason you are not satisfied with the book(s) you order, simply return it (them) within 15 days and receive a full refund. **BC**